Statistical and Probabilistic Models for Document Watermark Extraction

Kenu Rivera

TABLE OF CONTENTS

1	**Introduction**		
	1.1	Preamble	1
	1.2	Emerging Challenges in Digital Document Security	2
	1.3	Limitations of Human and Conventional Verification Methods	2
	1.4	Background, Classifications and Applications of Text Watermarking	3
	1.5	Motivation from Real-World Document Authentication	9
	1.6	Practical Importance of Watermark Text Extraction	9
	1.7	Importance of Automated Watermark Text Extraction	10
		1.7.1 Document Authentication	10
		1.7.2 Copyright Enforcement	10
		1.7.3 Anti-Counterfeiting	11
		1.7.4 Digital Forensics	11
		1.7.5 Document Retrieval and Classification	11
	1.8	Key Challenges in Watermark Extraction	12
		1.8.1 Low Intensity and Transparency	12
		1.8.2 Multi-Oriented Watermark Placement	12
		1.8.3 Overlapping with Printed Text	12
		1.8.4 Noise, Degradation, and Scanning Artifacts	13
		1.8.5 Variation in Font, Style, Opacity	13
		1.8.6 Lack of Robust Computational Models	13
	1.9	Limitations of Existing Manual and Conventional Approaches	15
	1.10	Motivation for a Multi-Stage Computational Framework	15
	1.11	Need and Motivation for This Research	16

1.12 Ethical, Legal, and Forensic Importance of Watermark Extraction 16
1.13 Problem Statement 17
1.14 Contribution to the Research Community 17
1.15 Research Objectives 17
1.16 Research Contributions 18
1.17 Summary 19
1.18 Organization of the Thesis 20

2. Literature Survey

2.1 Introduction 23
2.2 Text Watermarking and Extraction Techniques 23
2.3 Enhancement Techniques for Watermark Visibility 27
2.4 Clustering-Based Methods for Watermark Segmentation 28
2.5 Gaussian Mixture Models (GMM) in Document Segmentation 29
2.6 Markov Random Fields (MRF) for Spatial Modeling 31
2.7 Statistical and Machine Learning-Based Watermark Extraction 32
2.8 OCR-Based Evaluation of Extracted Watermark Text 33
2.9 Integrated Review: What Existing Methods Lack 34
2.10 Additional Literature on Watermarking, Image Enhancement and Document Processing 35
2.11 Document Segmentation and Binarization Literature 36
2.12 Literature on Texture and Background Modeling 37
2.13 Multi-Resolution and Frequency-Based Filtering Literature 38
2.14 Literature on Document Noise Removal and Restoration 38
2.15 Literature on Segmentation using Probabilistic Graphical Models 39
2.16 Literature on Handwriting and Stroke Extraction (Relevant to Curvature Preservation) 40

	2.17	Additional Works on Document Restoration, Watermark Extraction and Pixel-Level Modeling	41
	2.18	Multi-Oriented Text Detection	42
	2.19	Literature on Deep Learning Approaches (For Comparison)	44
	2.20	Recent Deep Learning-Based Watermarking and Detection Approaches	45
	2.21	Research Gap Analysis (Major Section)	46
	2.22	Extended Discussion on Research Gaps and Emerging Trends	49
	2.23	Summary	50
3.	**Watermark Enhancement using Max–Min Clustering**		
	3.1	Introduction	52
	3.2	Motivation for Watermark Enhancement	53
		3.2.1 Low Contrast between Watermark and Background	53
		3.2.2 Overlapping of Printed Text and Watermark	53
		3.2.3 Nonuniform Watermark Intensity	53
		3.2.4 Weak Edges and Faint Strokes	53
	3.3	Overview of Max–Min Clustering	53
	3.4	Proposed Methodology	54
		3.4.1 Max-Min Filtering Approach for Watermark Text Enrichment	54
		3.4.2 Mathematical Formulation of Max–Min Clustering	55
		3.4.3 Grouping of Neighborhood pixels for Watermark Text Detection	56
		3.4.4 Heuristic rules for Watermark Text Separation	57
	3.5	Dataset Collection and Description	58
	3.6	Experimental Analysis	59
	3.7	Sample Outcomes of the Proposed Model	61
	3.8	Comparative Analysis with Proposed Methodology	62
	3.9	Summary	63

4.	**Watermark Detection using Gaussian Mixture Modeling (GMM)**		
	4.1	Introduction	65
	4.2	Motivation for GMM-Based Detection	65
		4.2.1 Overlap in Intensity Distributions	65
		4.2.2 Local Nonuniformity	65
		4.2.3 Transparency and Blending	66
		4.2.4 Absence of Shape Cues	66
	4.3	Proposed Methodology	66
		4.3.1 Workflow of the Proposed Methodology	67
		4.3.2 Gaussian Mixture Model	68
		4.3.3 Probabilistic Modeling Using Gaussian Mixture Model	69
		4.3.4 Canny Edge Detector for Text Edge Components	70
		4.3.5 Morphological Operations and Otsu Thresholding	71
	4.4	Advantages of GMM for Faint Watermark Detection	73
	4.5	Experimental Analysis	73
		4.5.1 Metrics used for Evaluation of the proposed Model	77
	4.6	Sample Outcomes of the Proposed Model	77
	4.7	Comparative analysis with proposed model	79
	4.8	Summary	81
5.	**Watermark Extraction using Markov Random Field (MRF)**		
	5.1	Introduction	82
	5.2	Motivation for using MRF in Watermark Extraction	82
		5.2.1 Pixelwise Noise and False Positives	82
		5.2.2 Faint Curvature and Stroke Continuity	83
		5.2.3 Local Similarity	83
		5.2.4 Robustness to Scanner Noise and Illumination Variation	83
	5.3	Proposed Methodology	83

		5.3.1 K-Means Algorithm	84
		5.3.2 Canny Edge Detector	85
		5.3.3 Markov Random Field (MRF) for Watermark Text Extraction	87
		5.3.4 Markov Random Field Energy Function Derivation	89
	5.4	Experimental Analysis	89
		5.4.1 Performance Measures of The Proposed Model	90
	5.5	Sample Outcomes of the Proposed Model	91
	5.6	Comparative Analysis with Proposed Model	95
	5.7	Summary	96
6.	**OCR-Based Textual Evaluation in the Document**		
	6.1	Introduction	98
	6.2	Motivation for Refinement and Reconstruction	98
		6.2.1 Stroke Fragmentation after Extraction	98
		6.2.2 Geometry Distortion due to Printing and Scanning	99
		6.2.3 Artifact Removal	99
		6.2.4 Preservation of Fine Curvature	99
	6.3	Curvature-Based Boundary Refinement	99
	6.4	Summary	101
7.	**Conclusion and Future Work**		**103**
	Appendix		104
	Publication Details		108
	Sample Datasets of Text Watermark Images		
	1.	**Data set collected from Google**	109
	2.	**Public dataset**	115

CHAPTER 1

INTRODUCTION

1.1 Preamble

The exponential growth of digital infrastructure and the rapid transition from conventional paper-based documentation to electronic documents have fundamentally transformed the global information landscape. Government departments, academic institutions, financial organizations, healthcare units and private corporations increasingly prefer digital documents due to their convenience, cost-efficiency, speed of transmission and ease of storage [1]. However, this digital shift has simultaneously elevated concerns related to data integrity, authenticity and security. Unauthorized duplication, fraudulent manipulation, illegal circulation and alteration of digital documents have become significant cybersecurity challenges [2].

In response to these challenges, digital watermarking has emerged as a powerful technological solution for ensuring document authenticity and preventing unauthorized use. Watermarking embeds authentication information directly into the document, allowing verification of originality and protection against piracy. Although significant advancements have been made in multimedia watermarking particularly in images, audio and video, the domain of **text document watermarking remains underexplored** [3].

Visible watermark text is commonly used in certificates, government circulars, marksheets, legal contracts, tender documents and classification reports. The purpose of these watermarks is to demonstrate ownership, prevent misuse, discourage forgery, and retain credibility. However, real-world watermark text often appears faint, semi-transparent, multi-oriented (diagonal) and blended into the background. This makes extraction highly challenging using traditional image processing methods [4].

The extraction of watermark text plays a crucial role in several tasks, including document verification, security auditing, legal dispute resolution, digital forensics and automated document management. Manual verification is inefficient, subjective and prone to errors, especially for large-scale document processing systems. Therefore, there is a need for **robust computational models** capable of enhancing faint watermarks, detecting their boundaries accurately and extracting them without distorting the original content.

This dissertation aims to address the research gaps in watermark extraction by proposing an integrated computational framework combining enhancement, detection, segmentation and spatial modeling techniques. The proposed models employ Max-Min clustering, Gaussian Mixture Models (GMM), Markov Random Fields (MRF) and curvature-preserving refinement to deliver accurate and reliable watermark extraction from challenging document images.

1.2 Emerging Challenges in Digital Document Security

The rapid growth of e-governance, digital education platforms, online banking and electronic record management systems has significantly increased the circulation of digital documents across unsecured networks. While digital transformation improves accessibility and efficiency, it also exposes documents to unauthorized duplication, forgery and tampering. Official documents such as certificates, contracts, academic transcripts and identity records are particularly vulnerable, as even minor alterations can result in severe legal, financial and administrative consequences.

Visible watermarking is widely adopted as a deterrent against document misuse; however, its effectiveness depends not only on embedding but also on the ability to reliably extract and verify the watermark when disputes arise. In many real-world scenarios, watermark extraction is required after printing, scanning, compression or partial degradation of documents, making the extraction process far more challenging than ideal laboratory conditions.

1.3 Limitations of Human and Conventional Verification Methods

Manual inspection of watermarked documents is subjective, time-consuming, and error-prone, particularly when watermark text is faint or partially obscured by printed content. Human verification becomes infeasible when processing large document repositories such as government archives, educational records and financial databases. Traditional image processing techniques including global thresholding, morphological filtering and basic segmentation often fail to isolate watermark text due to low contrast, transparency and overlapping layers.

Furthermore, many existing systems rely on predefined rules or fixed thresholds, which lack adaptability to variations in watermark style, font, orientation and background texture. These limitations emphasize the necessity for automated and adaptive computational models capable of handling diverse document conditions with minimal human intervention.

1.4 Background, Classifications and Applications of Text Watermarking

Background of Text Watermarking:

Digital watermarking is a security technique that embeds ownership or authentication information within a multimedia object. Depending on the visibility and robustness, watermarks are classified as follows.

- **Visible watermarks**

Visible watermarks are watermarking techniques in which identification information is intentionally embedded in a form that is clearly noticeable to the viewer, typically as logos, symbols or textual overlays placed on the document image. These watermarks are commonly used to indicate ownership, copyright or document status and are often applied with partial transparency or repeated patterns to prevent easy reuse of the document. While visible watermarks are effective as an immediate visual deterrent against unauthorized copying and misuse, they can interfere with document readability and overall appearance. In addition, because the watermark is openly present, it can be removed or altered using image editing tools, which limits its suitability for applications requiring strong security, authentication or forensic verification.

- **Invisible watermarks**

Invisible watermarks are a form of document watermarking in which identifying information is hidden within the document image without producing any noticeable visual change. This is achieved by making carefully controlled statistical or structural alterations to the document such as modifying text geometry, spatial relationships or underlying image characteristics in a manner that remains imperceptible to human readers. Since the watermark does not affect the visible appearance of the document, it preserves readability and layout while still allowing the embedded information to be detected or extracted using specialized algorithms. These watermarks are particularly effective for authentication and copyright protection, as their hidden nature makes them difficult to identify or remove through casual inspection or simple editing operations.

- **Fragile watermarks**

Fragile watermarks are watermarking techniques designed to be extremely sensitive to changes in the document image, such that even minor modifications can disrupt or destroy the embedded watermark. The primary purpose of fragile watermarking is not robustness but integrity verification, as any form of tampering such as editing text, altering pixels or applying basic image processing operations can be detected through watermark degradation or loss. Because these watermarks do not survive document modifications, they are well suited for applications like authenticity verification and tamper detection, where the absence or alteration of the watermark serves as clear evidence that the document has been modified.

- **Robust watermarks**

Robust watermarks are designed to remain detectable even when a document image undergoes common image processing operations or intentional attacks. These watermarks are embedded in a manner that allows them to survive modifications such as compression, noise addition, resizing, rotation and printing-scanning processes. The emphasis of robust watermarking is on durability rather than sensitivity, enabling reliable extraction of the embedded information despite changes to the document's visual representation. As a result, robust watermarks are widely used for copyright protection, ownership verification and forensic tracking in environments where documents are frequently transformed or reproduced.

Text watermarking is classified into:

- **Linguistic watermarking**

Linguistic watermarking is a technique that embeds watermark information by altering the language structure of textual content rather than modifying its visual appearance. This approach relies on controlled changes such as reordering sentence components, substituting words with semantically equivalent alternatives or adjusting syntactic constructions while preserving the original meaning of the text. Because the watermark is encoded at the linguistic level, it does not affect the visual layout of the document and remains intact even when the text is copied or reformatted. Linguistic watermarking is particularly suitable for electronic text documents; however, its application to document images is limited as such methods require reliable text interpretation and may lose effectiveness after scanning or image-based transformations.

- **Structural watermarking**

Structural watermarking is a technique in which watermark information is embedded by making subtle modifications to the physical structure of text within a document [5]. This includes controlled adjustments to line spacing, spacing between characters or words and slight variations in font shape or character geometry. These changes are designed to be small enough to avoid affecting readability while still allowing the embedded information to be detected during extraction. By operating on the layout and structural properties of text rather than pixel intensity values, structural watermarking can achieve a balance between imperceptibility and robustness particularly in text-dominated document images.

- **Image-based text watermarking**

Image-based text watermarking is an approach in which a document is handled purely as an image rather than as structured textual content. In this method, watermark information is embedded by introducing carefully controlled patterns or modifications at the pixel level, typically in regions containing text or background areas. The embedding process does not rely on character recognition or linguistic analysis which makes it applicable to a wide range of document types and scripts. However, because the watermark is tied to pixel-level properties, its effectiveness and extractability can be affected by common image processing operations such as compression, noise addition, resizing or scanning.

Types of Watermarking Based on Content and Placement:

Text watermarking can further be categorized based on the **number of watermarks, document components and content type** as described below.

- **Single Watermarking**

Single watermarking is a watermarking approach in which only one watermark is embedded within a document to convey essential information such as ownership, authorship or authenticity. The embedded watermark serves a specific purpose and is typically designed to be easily detectable or extractable using a defined procedure. Because only a single watermark is used as this approach simplifies the embedding and extraction processes and reduces the risk of visual distortion or interference between multiple watermarks. Single watermarking is commonly adopted in applications where straightforward copyright identification or document verification is sufficient.

- **Multiple Watermarking**

Multiple watermarking is a technique in which two or more watermarks are embedded within a single document to provide layered protection and enhanced security. Each watermark can serve a distinct purpose such as indicating ownership, verifying authenticity or enabling traceability of the document through different distribution channels. By embedding multiple watermarks, the system increases resilience against tampering, forgery or unauthorized removal, since altering one watermark does not necessarily compromise the others. This approach is particularly useful in sensitive applications where multi-level security and robust authentication are required, although it requires careful design to avoid visual degradation or interference between watermarks.

- **Foreground Watermarking**

Foreground watermarking is a method in which watermark information is embedded directly into the visible elements of a document such as text characters, logos or graphical content. By altering the foreground features, the watermark becomes closely integrated with the main content making it more apparent for ownership verification or authenticity checks. This approach allows the watermark to interact with the document's primary information while maintaining readability and visual quality, but it may be more susceptible to removal or distortion if the foreground elements are edited or modified.

- **Background Watermarking**

Background watermarking is a technique in which watermark information is embedded into the background areas of a document rather than its main textual or graphical content. This is often applied to elements such as blank spaces, patterned fills or subtle textures, making the watermark less intrusive while remaining detectable. Background watermarking is widely used in official documents, certificates and legal papers, where maintaining the readability and appearance of the main content is crucial. Because the watermark is integrated into the background, it is more difficult to remove without affecting the overall document, providing an additional layer of security and authenticity verification.

- **Image Watermarking in Text Documents**

Image watermarking in text documents is a technique where watermark information is embedded specifically into the visual elements contained within the document, such as images, logos, seals or signatures. Instead of altering the textual content, the watermark is integrated into these graphical components, allowing the main text to remain unchanged and fully readable. This approach is particularly useful for protecting documents that include official symbols or illustrations, as it ensures the embedded watermark can be detected for authentication or ownership verification while preserving the visual integrity of the document's images.

- **Handwritten Watermarking**

Handwritten watermarking is a technique in which watermark information is embedded into handwritten elements of a document, such as signatures, notes or annotations. This method is often applied in legal, administrative and official records where handwritten content plays a critical role in authentication and validation. By integrating the watermark into the unique characteristics of handwriting such as stroke patterns or pressure variations, the watermark remains inconspicuous while enabling verification of document authenticity and protection against forgery or tampering.

- **Printed and Scanned Document Watermarking**

Printed and scanned document watermarking refers to techniques in which watermark information is embedded into a document in a way that allows it to survive the physical printing and subsequent scanning processes. These methods ensure that the watermark remains detectable even after the document undergoes transformations such as resolution changes, color alterations, noise introduction or slight distortions that commonly occur during printing and scanning. This type of watermarking is particularly important for official or archival documents, where maintaining both the visual integrity of the printed material and the ability to verify authenticity after scanning is essential.

Applications of Image-Based Text Watermarking

Among the various techniques, **image-based text watermarking** is the most widely adopted due to its applicability to real-world documents such as:

- Government certificates - Such as birth certificates, passports, voter ID cards and educational certificates.
- Administrative orders - Such as official memos, circulars and notifications.
- Educational certificates - Such as diplomas, degree certificates and transcripts.
- Bank statements - Such as account summaries and transaction reports.
- Property documents - Such as land deeds, sale agreements and title certificates.
- Identification cards - Such as passports, driver's licenses and employee ID cards.
- Scanned and printed official records - Such as government orders, legal documents and archival records.

However, although embedding watermarks in such documents is relatively straightforward, **watermark extraction and detection especially when watermarks are faint, blended or degraded due to printing, scanning or noise remain significantly more challenging**. This challenge necessitates the development of robust watermark extraction and detection techniques.

1.5 Motivation from Real-World Document Authentication

With the widespread adoption of digital documentation across governmental, educational, financial and legal domains, ensuring document authenticity has become a critical concern. Documents such as academic certificates, bank statements, legal affidavits, property records and identity proofs are frequently exchanged in digital or scanned formats. To prevent unauthorized duplication and misuse, visible text watermarks are commonly embedded in these documents. These watermarks often carry ownership information, institutional identifiers or confidentiality notices and are intentionally designed to be semi-transparent so that they do not obstruct the main content.

In practical scenarios, watermark extraction is often required during dispute resolution, forensic investigation or authenticity verification. However, documents are rarely available in their original digital form; instead, they undergo print-scan processes, compression, photocopying and storage-related degradation. These operations significantly reduce watermark visibility, introduce noise and cause overlapping between watermark text and foreground content. Consequently, extracting watermark text reliably under such conditions remains a challenging task, motivating the need for robust and automated extraction techniques.

1.6 Practical Importance of Watermark Text Extraction

Reliable extraction of watermark text is crucial in several practical applications. In legal and administrative environments, extracted watermarks serve as evidence for document authenticity and ownership verification. In academic institutions, watermark extraction assists in validating certificates and preventing credential fraud. Financial organizations rely on watermark verification to ensure the integrity of transaction records, statements and agreements. Additionally, in digital forensics and cybercrime investigations, watermark text extraction provides critical metadata for tracing document origin and detecting unauthorized modifications.

With the increasing use of scanned documents and portable document formats (PDFs), watermark extraction systems must operate effectively across varying resolutions, illumination conditions and noise levels. This real-world complexity necessitates robust extraction frameworks that go beyond conventional enhancement or segmentation techniques.

1.7 Importance of Automated Watermark Text Extraction

Watermark text extraction plays a crucial role in multiple domains. Automated watermark text extraction is a critical component in modern document security systems, as it allows the embedded watermark information to be detected and retrieved without manual intervention. This capability is essential for ensuring the authenticity, ownership and integrity of digital and printed documents across various domains. In academic and research environments, automated extraction helps verify authorship and prevent plagiarism. In legal and administrative contexts, it provides a reliable mechanism to detect tampering or forgery in contracts, certificates and official records. In corporate and commercial applications, it supports copyright protection, traceability of confidential documents and secure distribution of sensitive information. By automating the extraction process, organizations can handle large volumes of documents efficiently, reduce human error and enforce security measures consistently, making automated watermark extraction a vital tool for maintaining trust and accountability in document management systems.

1.7.1 Document Authentication

Watermark extraction is an essential process for verifying the authenticity of documents [6]. By retrieving the embedded watermark, it can be confirmed whether a document originates from a recognized or trusted institution. This process helps ensure that certificates, legal papers or official records are genuine and have not been forged. Additionally, watermark extraction can detect any alterations or tampering that may have occurred after the document was issued. It provides a reliable mechanism to maintain the integrity and credibility of important documents. Overall, automated extraction enhances security and trust in document management systems.

1.7.2 Copyright Enforcement

Watermarks are embedded in documents to establish clear ownership and protect intellectual property. By extracting the watermark, organizations can demonstrate that the content belongs to them, helping to prevent unauthorized copying, plagiarism or illegal distribution. This process ensures that original authors or institutions maintain control over their work and can take appropriate action against misuse. Automated watermark extraction makes it easier to monitor and enforce copyright, especially when documents are shared digitally or circulated widely.

1.7.3 Anti-Counterfeiting

Watermarks are used in important documents such as certificates, agreements, identity cards and academic marksheets to prevent unauthorized duplication and fraudulent activities. By embedding hidden or visible watermark information, the authenticity of the document can be verified, making it difficult for counterfeit copies to pass as genuine. Extraction of these watermarks allows institutions and authorities to detect tampering or forgery, ensuring that only legitimate documents are accepted and trusted. This technique plays a crucial role in maintaining the integrity and security of official records.

1.7.4 Digital Forensics

Extracted watermarks serve as important evidence in digital forensics by helping to trace the origin and ownership of documents. They are useful in investigations of cybercrime, intellectual property disputes and cases of document tampering. By analyzing the embedded watermark, authorities can determine whether a document has been altered, identify the responsible parties and support legal or cybersecurity actions. This makes watermark extraction a valuable tool for maintaining accountability and integrity in digital document management.

1.7.5 Document Retrieval and Classification

Embedded watermarks can store metadata like department name, year of creation, classification level or document type, facilitating efficient organization of documents. By extracting this information, automated systems can quickly retrieve, sort and categorize documents without manual intervention. This improves document management, reduces search time and ensures that sensitive or classified information is handled appropriately. Watermark-based retrieval and classification also help maintain consistency and accuracy in large-scale digital archives.

Given the increasing use of scanned documents and PDFs in official workflows, the need for **automated, reliable watermark extraction systems** has become more important than ever.

1.8 Key Challenges in Watermark Extraction

Extracting watermark text from real-world documents remains difficult due to multiple factors:

1.8.1 Low Intensity and Transparency

Watermarks are often designed to be subtle or semi-transparent so that they do not interfere with the readability or appearance of the document [7]. However, this faintness makes the watermark difficult to detect and extract, as its edges are less defined and its contrast with the surrounding text or background is minimal. Low intensity and transparency increase the likelihood of errors during extraction, especially when the document undergoes scanning, compression or other processing that can further diminish the watermark's visibility.

1.8.2 Multi-Oriented Watermark Placement

To improve security, watermarks are often placed in multiple orientations, such as diagonally, across pages or at varying angles [8]. While this makes unauthorized removal or forgery more difficult, it also complicates the extraction process. Automated extraction systems must accurately detect and interpret watermarks regardless of their orientation, which increases computational complexity and the likelihood of errors, especially in documents with dense text or complex layouts.

1.8.3 Overlapping with Printed Text

In many documents, watermarks are embedded directly beneath or within the areas containing printed text. This overlap can obscure the watermark's boundaries and reduce its distinguishability from the main content, making extraction challenging. When the watermark and printed text interfere visually, automated systems may struggle to separate them accurately, leading to incomplete or incorrect recovery of the embedded information.

1.8.4 Noise, Degradation, and Scanning Artifacts

Watermark extraction is often hindered by factors such as low-resolution scanning, uneven illumination, shadows or the natural texture of the paper [9]. These conditions introduce noise and degrade the document image, reducing the clarity and visibility of the embedded watermark. As a result, automated extraction systems may face difficulties in accurately detecting or reconstructing the watermark, leading to errors or incomplete recovery.

1.8.5 Variation in Font, Style, Opacity

Watermarks can vary widely in terms of font type, size, thickness, color, transparency and overall design across different documents. These variations make automated extraction more difficult, as the system must adapt to diverse visual characteristics while maintaining accuracy. Differences in style and opacity can reduce contrast with the background or printed content, increasing the risk of partial detection or extraction errors.

1.8.6 Lack of Robust Computational Models

Current watermark extraction techniques often struggle to handle the wide diversity of watermark designs and document formats. Many models are tailored to specific fonts, layouts or embedding methods, limiting their ability to generalize across different types of documents. This lack of robustness results in decreased extraction accuracy when faced with new or complex watermark styles, making it challenging to develop universally reliable extraction systems.

These challenges necessitate advanced computational solutions beyond simple thresholding or segmentation.

Figure 1.1 illustrates different watermarked document images.

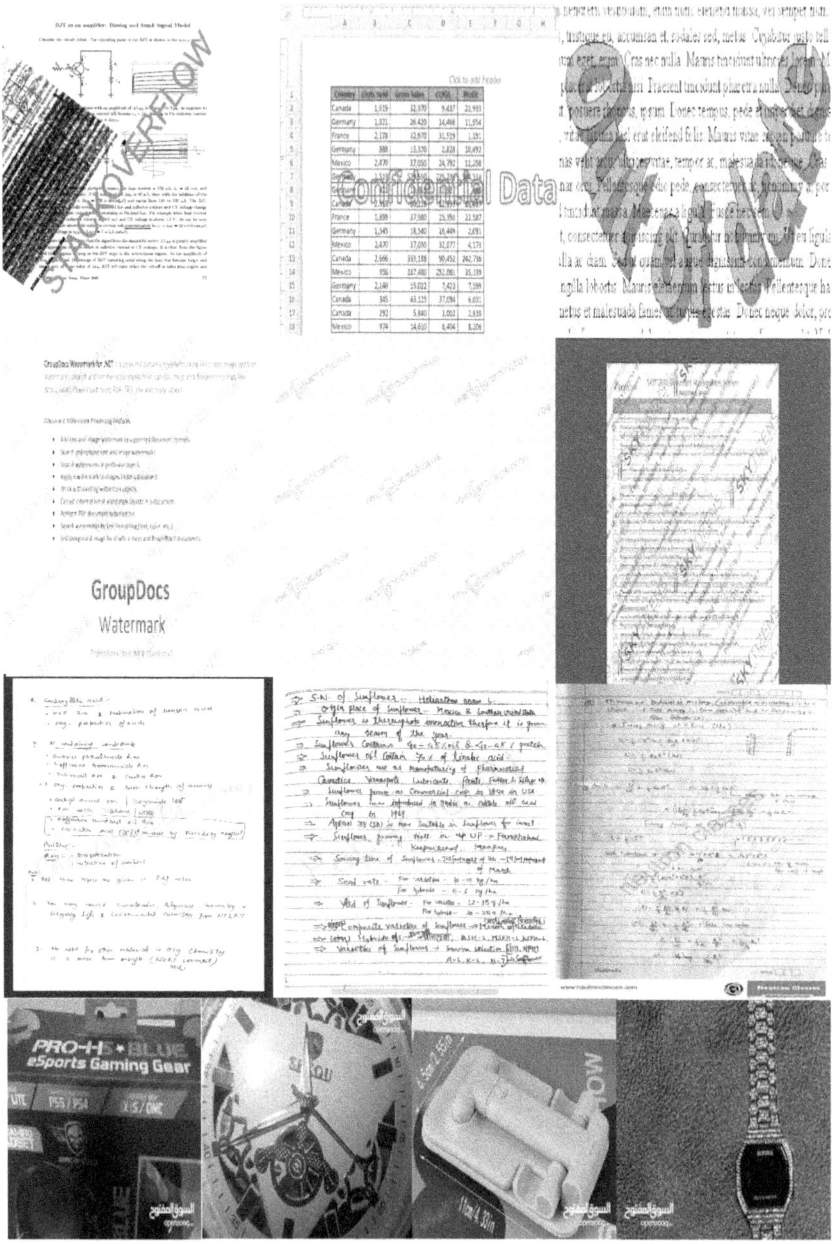

Figure 1.1: Sample Data Set Images

1.9 Limitations of Existing Manual and Conventional Approaches

Manual verification of watermarked documents relies heavily on human perception and experience, making it subjective and inconsistent. Faint or low-contrast watermark text is often overlooked during visual inspection, particularly when documents contain dense textual or graphical content. Moreover, manual verification is time-consuming and impractical for large-scale applications such as institutional record verification, recruitment processes and archival management systems.

Traditional image processing approaches, including thresholding, edge detection and morphological operations, have been widely explored for watermark extraction. While these techniques perform adequately under controlled conditions, they are highly sensitive to illumination changes, noise and background variations. Fixed threshold values and rule-based operations fail to generalize across different watermark styles, font sizes and document layouts. These limitations highlight the necessity for adaptive, statistically driven and spatially consistent extraction frameworks capable of handling real-world document variability.

1.10 Motivation for a Multi-Stage Computational Framework

Existing watermark extraction methods often address isolated components of the problem, such as enhancement or detection, without considering their interdependence. Enhancement techniques may improve visibility but distort boundaries, while segmentation methods may detect regions but fail to preserve stroke continuity. Similarly, statistical models may classify pixels accurately but ignore spatial coherence, resulting in fragmented watermark text.

To overcome these challenges, there is a strong motivation to design a unified, multi-stage framework that integrates enhancement, statistical detection, spatial modeling and boundary refinement. Such a framework should be capable of adapting to faint, semi-transparent and multi-oriented watermark text while preserving structural details essential for readability and OCR-based evaluation.

1.11 Need and Motivation for this Research

Existing watermark extraction methods suffer from:

- Weak performance on faint or semi-transparent watermarks
- Inability to retain fine boundary structures
- Failure under noisy or degraded conditions
- Poor performance for diagonal or rotated watermarks
- Insufficient spatial modeling
- Lack of integrated pipelines for enhancement + detection + extraction

This motivates the development of a **multi-stage framework** integrating:

- Max-Min clustering for enhancement
- Gaussian Mixture Modeling for detection
- Markov Random Field (MRF) for spatially consistent extraction
- Curvature analysis and Laplacian of Gaussian (LoG) for boundary refinement

Such a combined approach has not been adequately explored in the literature, making this research both relevant and novel.

1.12 Ethical, Legal, and Forensic Importance of Watermark Extraction

Watermark text extraction plays a crucial role in digital forensics and legal document analysis. Extracted watermarks often serve as evidential proof for establishing document ownership, detecting unauthorized reproductions and resolving copyright disputes. In judicial proceedings, reliable watermark extraction enhances the credibility of document evidence and supports objective decision-making.

From an ethical standpoint, automated watermark extraction systems contribute to fraud prevention, identity protection and intellectual property enforcement. By reducing reliance on subjective human judgment, such systems promote transparency, consistency and fairness in document verification processes. Therefore, watermark text extraction is not merely a technical problem but also a socially and legally significant research domain.

1.13 Problem Statement

Real-world watermark text extraction is hindered by low visibility, noise, structural overlap and lack of robust computational techniques.

Problem Statement:

To develop an efficient and accurate computational framework for enhancing, detecting and extracting watermark text from document images while preserving boundary clarity and minimizing distortion, despite variations in opacity, orientation, noise and document quality.

1.14 Contribution to the Research Community

This research contributes to the field of document image analysis by addressing the less explored yet practically critical problem of visible watermark text extraction. By combining classical statistical modeling with spatial dependency analysis and curvature-aware refinement, the proposed approach offers an interpretable and computationally efficient alternative to data-hungry deep learning models. The framework is designed to operate effectively on small and diverse datasets, making it suitable for real-world deployment in document authentication and security systems.

1.15 Research Objectives

1. The implementation of an effective method for enhancement in text watermark.
2. Development of boundary detection to preserve the information in watermarking.
3. A novel approach for detecting the watermark text present in images.
4. The implementation of an efficient model for watermark text extraction.
5. To evaluate extracted watermark quality using OCR-based metrics.

1.16 Research Contributions

- A novel enhancement model combining intensity-based clustering and local contrast maximization.
- A GMM-driven detection mechanism identifying watermark clusters.
- A robust MRF-MAP-based extraction algorithm preserving spatial dependency.
- A curvature-based boundary refinement technique.
- A complete OCR evaluation pipeline for watermark extract quality assessment.
- A comprehensive comparison with existing methods demonstrating superior performance.

Figure1.2 shows overall architecture of the proposed watermark extraction from document images

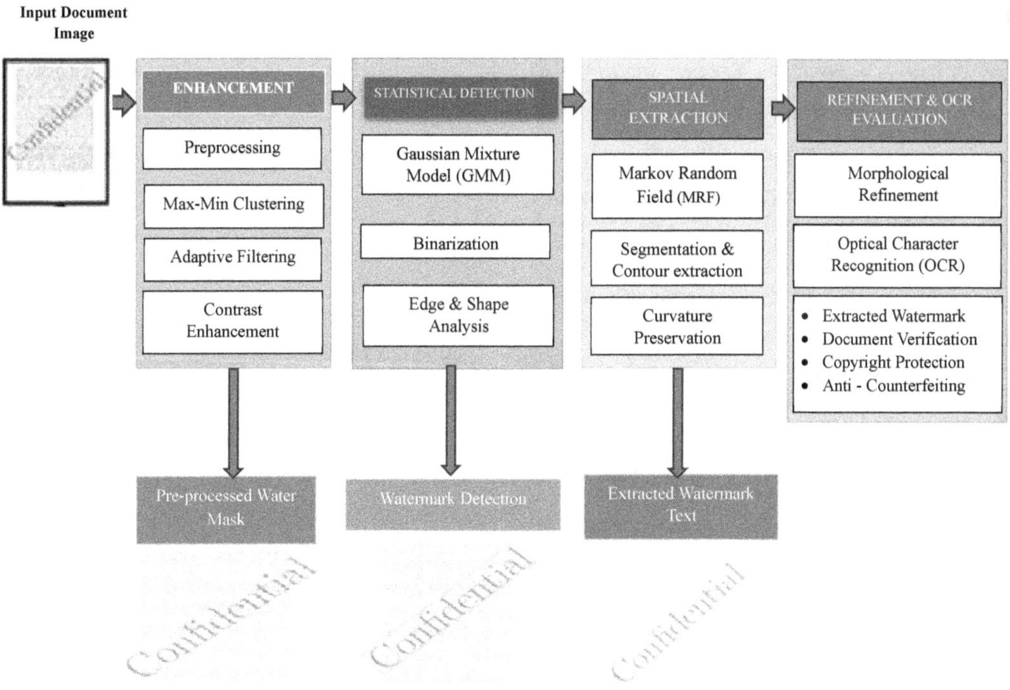

Figure 1.2: Overall Architecture of the Proposed Watermark Extraction Framework

1.17 Summary

Chapter 1 introduced the fundamental motivation behind watermark text extraction from digital document images in the context of increasing reliance on electronic documentation. The chapter highlighted the growing challenges related to document authenticity, copyright protection and tamper detection arising from unauthorized duplication and manipulation of digital documents. It emphasized that although visible watermarks are widely used in official, legal and academic documents, extracting such watermarks is highly challenging due to low contrast, transparency, multi-orientation, overlap with printed text and scanning artifacts.

The chapter discussed the limitations of existing watermark extraction techniques and identified critical research gaps, including weak performance on faint watermarks, lack of boundary preservation, insufficient spatial modeling and absence of integrated end-to-end extraction frameworks. Based on these observations, the problem statement was formulated and clear research objectives and contributions were outlined. Finally, the organization of the thesis was presented, providing a structured overview of the proposed multi-stage watermark extraction framework developed in subsequent chapters.

1.18 Organization of the Thesis

This dissertation is organized into seven chapters, each addressing a critical component of the research in a systematic and progressive manner. The structure of the thesis ensures clarity, logical flow and coherence between theoretical foundations, methodological development, experimental validation and concluding insights.

Chapter 1 - Introduction

The first chapter presents the background and motivation for watermark text extraction in digital documents. It discusses the increasing reliance on digital documentation and the consequent need for reliable authentication and copyright protection mechanisms. The chapter highlights the major challenges associated with extracting faint, overlapping and multi-oriented watermark text from document images. It formulates the research problem, outlines the objectives, presents the significance and scope of the study, specifies the contributions made and concludes with an overview of the thesis structure.

Chapter 2 - Literature Survey

Chapter 2 provides an extensive and in-depth review of the existing literature related to digital watermarking, text watermarking, document image processing, clustering techniques, Gaussian Mixture Models (GMM), Markov Random Fields (MRF) and curvature-based edge detection methods. Each significant publication is discussed in detail, examining its methodology, strengths, limitations and relevance to the present work. The chapter also highlights the research gaps in current watermark extraction techniques, which serve as the foundation for developing the proposed models. This chapter is comprehensive and positions the proposed research within the broader context of existing scholarship.

Chapter 3 -Watermark Enhancement using Max–Min Clustering

This chapter introduces the first major component of the proposed methodology: watermark enhancement. It explains the rationale behind using Max–Min clustering to increase the intensity gap between watermark and background pixels. The chapter describes the mathematical formulation, algorithmic steps, enhancement strategy, and implementation details. Experimental results demonstrating improved watermark visibility compared to existing methods are also presented.

Chapter 4 -Watermark Detection using Gaussian Mixture Modeling (GMM)

Chapter 4 presents the second component of the proposed framework: watermark detection. The chapter explains how GMM is used to statistically model document pixel distributions and identify watermark clusters. The process of parameter estimation through Expectation-Maximization (EM), cluster likelihood computation and separation of watermark regions is described in detail. The performance of GMM-based detection is validated through several experiments, demonstrating its effectiveness in distinguishing faint watermark pixels from background and printed text.

Chapter 5 - Watermark Extraction using Markov Random Fields (MRF)

Chapter 5 introduces the primary extraction model using MRF combined with Maximum A Posteriori (MAP) estimation. The chapter elaborates on spatial dependency modeling, energy function design, clique potential formulation and label optimization. It presents how MRF overcomes the limitations of GMM by enforcing spatial consistency, resulting in clean and continuous watermark strokes. The chapter also includes algorithmic comparisons with existing segmentation techniques.

Chapter 6 - Curvature-Based Refinement and OCR Evaluation

This chapter discusses the final refinement process using curvature structure analysis and Laplacian of Gaussian (LoG) filtering to preserve fine edges and enhance boundary clarity of the extracted watermark text. The refined watermark output is evaluated using Optical Character Recognition (OCR) metrics to assess readability and structural preservation. Quantitative and qualitative evaluations are presented, showing how the proposed multi-stage framework improves recognition accuracy compared to existing approaches.

Chapter 7 - Conclusion and Future Work

The final chapter summarizes the key findings of the research, re-emphasizes the contributions made and discusses the implications of the proposed methodology. It outlines the limitations encountered during the study and provides several directions for future research, including potential improvements using deep learning, hybrid models and advanced probabilistic techniques. The chapter concludes by emphasizing the significance of the work in advancing watermark extraction and document authentication technologies.

CHAPTER 2

LITERATURE SURVEY

2.1 Introduction

The purpose of this chapter is to present a comprehensive literature review covering all major research areas related to the extraction of watermark text from document images. The studies reviewed in this chapter include foundational theories of digital watermarking, methods for embedding and extracting text-based watermarks, image enhancement strategies, clustering and segmentation models, Gaussian Mixture Models (GMM), Markov Random Fields (MRF), curvature-based edge detection and OCR-based evaluation systems. Since visible watermarks are widely used in administrative, academic, legal and financial documents, understanding prior works is essential for building an efficient extraction methodology.

Existing literature shows that watermark extraction is fundamentally more challenging than watermark embedding, mainly due to low contrast, transparency, overlapping with printed text, scanning noise and geometric distortions. The review reveals that most existing techniques lack robustness and generalization when applied to real-world degraded documents. Therefore, the literature survey not only summarizes previous contributions but also identifies critical research gaps that motivate the development of the proposed framework.

2.2 Text Watermarking and Extraction Techniques

Text watermarking is a specialized area of digital watermarking that focuses on embedding information within textual content or text-based document images to ensure authenticity, ownership and security. Unlike multimedia watermarking, text watermarking must preserve the readability and visual quality of the document, making it a challenging task. Techniques for text watermarking can be broadly classified into spatial-domain methods, which directly modify pixel values or text layout and feature-based methods, which exploit structural characteristics of text such as character shapes, spacing or line alignment. Extraction techniques are designed to retrieve the embedded information and they can be blind, semi-blind or non-blind depending on whether the original document is needed for recovery. Over the years, researchers have explored various approaches including pixel-level modifications, linguistic watermarking, structural alterations and transform-domain embedding. Extraction methods often combine image processing, pattern recognition and in some cases OCR-assisted analysis to recover watermarks reliably. Despite significant advancements, challenges such as document degradation, low contrast, variable fonts and complex layouts continue to limit the accuracy and robustness of watermark extraction, motivating ongoing research into more efficient and adaptive methods.

Brassil et al. (2002) - Visible and Invisible Text Watermarks

In this study, the researchers extended their earlier work by introducing mixed visible-invisible watermarking strategies suitable for printed and scanned documents. They demonstrated that visible watermarks (like large stamp-text overlays) are more practical for real-world document security. However, they also noted that extracting faint or semi-transparent visible watermarks is significantly more challenging than inserting them. This observation supports the need for advanced extraction techniques, particularly those involving statistical modeling like GMM and spatial dependency modeling like MRF [10].

Feng and Huang (2012) - DCT-based Zero-Watermarking Algorithms

In this study, the researchers proposed one of the early *DCT-based zero-watermarking algorithms* designed specifically for text images. Their method avoids embedding by extracting watermark features from DCT coefficients, making it suitable for fragile, low-distortion text documents. However, the method performs poorly on scanned documents where compression and noise degrade frequency-domain consistency [11].

Alattar and Alattar (2004) - Line Spacing and Paragraph Justification Modifications

In this study, the researchers introduced a structural text watermarking technique using line spacing and paragraph justification modifications. This linguistic-structural approach is effective for digitally typeset text but fails on scanned documents or images containing visible watermark overlays [12].

Maxemchuk and Low (1998) - Line-Shifting and Word-Spacing Techniques

In this study, the researchers performed a performance evaluation of two classical text marking methods, highlighting that line-shifting and word-spacing techniques are vulnerable to reflow, rescaling and low-quality scanning. This study motivated later research focusing on pixel-level watermark extraction instead of syntactic modifications [13].

Maxemchuk (1994) - Embedded Authentication Markers

In this study, the researchers also discussed electronic document distribution and the need for embedded authentication markers in printed and transmitted text documents, establishing the early motivation for watermarking-based authentication [14].

TamilSelvan et al. (2009) - Eigenvalue-based Watermarking Technique

In this study, the researchers proposed an *eigenvalue-based* watermarking technique for text documents where watermark information is embedded through linear algebraic transformations. While computationally strong, the method lacks robustness against geometric distortions [15].

Palit and Garain (2006) - Run-Length Encode

In this study, the researchers presented watermarking of *symbolically compressed documents* using run-length encoded pages. Although efficient, it applies only to compressed TIFF-like documents, not natural-image text documents where the watermark is faint and transparent [16].

Singh and Sharma (2021) - Zero-Watermarking in Wavelet Domain

In this study, the researchers developed a tamper-detection technique for document images using zero-watermarking in the wavelet domain. Their method relies on extracting watermark features without embedding, making it suitable for fragile documents. This aligns with watermark extraction scenarios where any embedding may distort the original document [17].

Nazari and Mehrabian (2021) - Chaotic IWT-LSB Watermarking

In this study, the researchers proposed a blind watermarking approach combining Integer Wavelet Transform (IWT) with chaotic LSB embedding. Although designed for medical images, their flexible-capacity embedding method demonstrates that chaotic maps improve watermark resilience against noise and compression conditions similar to faint watermark degradation in scanned document [18].

Gong, Qin, and Zhang (2021) - Visible Watermarking via Fuzzy Systems

In this study, the researchers introduced a visible watermarking technique for document images using a two-stage fuzzy inference system. Their model adjusts watermark visibility adaptively based on document background characteristics, which is relevant for watermark extraction since the detection difficulty depends on local contrast [19].

Zenati et al. (2021) - Steganography with YUV Document Embedding

In this study, the researchers presented a steganography scheme for hiding online signatures inside YUV document images. Although primarily a hiding technique, it demonstrates how document images can carry additional embedded structures, further reinforcing the need for robust extraction and detection frameworks [20].

Zear and Singh (2021) - Dual Watermarking with LWT-DCT-SVD

In this study, the researchers proposed a robust dual-watermarking approach using a hybrid LWT-DCT-SVD architecture. Their dual-domain method enhances robustness under compression and geometric distortions, which are also significant challenges in watermark extraction from degraded document images [21].

Barlaskar et al. (2022) - GA + DCNN + SVR Hybrid Extraction

In this study, the researchers introduced a genetic algorithm-optimized hybrid deep CNN and SVR framework for watermark extraction. Their method demonstrates the power of combining ML with statistical models for robust watermark recovery-similar to your work involving GMM-MRF modeling [22].

Singh, R., et al. (2023) - Lightweight Transformer for Real-Time Extraction

In this study, the researchers developed a real-time watermark extraction system using lightweight transformer networks. Their attention-based design shows that transformer architectures can effectively isolate faint watermark traces from document backgrounds with high speed and accuracy [23].

2.3 Enhancement Techniques for Watermark Visibility

Tomasi & Manduchi (1998) - Bilateral Filtering

In this study, the researchers introduced the bilateral filter, a nonlinear filter that smoothens images while preserving edges. This is relevant to watermark extraction because faint watermark boundaries can be easily blurred by traditional smoothing operations. Bilateral filtering maintains edge sharpness, making watermark structures more visible. However, bilateral filtering alone cannot completely isolate watermark text; thus, it is used in combination with clustering or segmentation [24].

Perona & Malik (1990) - Anisotropic Diffusion

In this study, the researchers introduced anisotropic diffusion, a Partial Differential Equation (PDE)-based method that enhances images by diffusing intensity along smooth regions while preserving edges. This technique is helpful for enhancing watermark text that is smooth and faint. Their method reduces noise without blurring high-frequency details such as watermark edges. The theory behind anisotropic diffusion influences the curvature processing used later in this thesis [25].

Marr & Hildreth (1980) - Laplacian of Gaussian (LoG)

In this study, the researchers proposed the LoG operator for edge detection, which enhances curvature structures in images. This makes LoG highly relevant to watermark extraction because watermark characters often consist of strokes with distinct curvature properties. LoG is particularly useful for detecting watermark boundaries in low-contrast images, which is why it forms part of the refinement phase in the proposed methodology [26].

Tumblin and Rushmeier (1993)

In this study, the researchers introduced tone reproduction methods for realistic images, emphasizing perceptual preservation of contrast. Their principles influence watermark enhancement because faint watermark strokes require tone-preserving filtering [27].

Durand and Dorsey (2002)

In this study, the researchers proposed *fast bilateral filtering* for HDR tone-mapping, maintaining local contrast while compressing intensity ranges. This supports watermark visibility improvement because bilateral filters help preserve edges of faint watermark strokes [28].

Maiorana and Campisi (2016) - Perceptually-aware HDR Watermarking

In this study, the researchers developed a *perceptually-aware HDR watermarking model.* Their multi-bit HDR watermarking approach shows that watermark visibility is strongly dependent on perceptual contrast, which aligns with the motivation for using curvature and LoG-based refinement [29].

Zhang et al. (2021) - High-Boost Filtering for Text Enhancement

In this study, the researchers proposed an adaptive high-boost filtering technique to enhance degraded text in documents. Their method improves stroke visibility under noise, blur and low-contrast conditions, which directly supports your motivation to use curvature, LoG and bilateral filters to enhance faint watermark regions before extraction [30].

2.4 Clustering - Based Methods for Watermark Segmentation

Clustering-based methods for watermark segmentation are widely used in text watermark extraction to separate the embedded watermark from the main content of a document image. These approaches group pixels or image features into clusters based on similarity measures such as intensity, color, texture or spatial location. By analyzing these clusters, it becomes possible to distinguish watermark regions from background text or graphical content, even when the watermark is faint or partially overlapping with printed text. Common clustering techniques include K-means, fuzzy C-means and hierarchical clustering, which iteratively partition image data to maximize intra-cluster similarity and minimize inter-cluster similarity. Clustering-based segmentation is particularly effective for detecting irregular or semi-transparent watermarks, as it does not rely on predefined shapes or templates. However, its accuracy can be influenced by noise, variations in document quality and low contrast between watermark and text. Overall, clustering provides a flexible and adaptive framework for isolating watermark features, serving as a foundational step for subsequent extraction and verification processes.

MacQueen (1967) - K-Means Clustering

MacQueen's K-Means clustering laid the foundation for unsupervised segmentation. K-means is widely used in watermark detection to separate background, foreground and watermark clusters based on pixel intensities. However, in documents with overlapping watermark and printed text intensities, K-means fails because of its hard assignment nature. This limitation led to the use of **Max-Min clustering** in the present work which increases the contrast between cluster centroids [31].

Bezdek (1981) - Fuzzy C-Means (FCM)

Bezdek introduced Fuzzy C-Means, which assigns soft membership values to clusters. FCM has been used for document segmentation, especially when pixel intensity boundaries are fuzzy. Although FCM performs better than K-Means for faint watermark detection, it is extremely sensitive to noise-a major problem in scanned documents. This study shows the need for robust clustering approaches capable of handling noise and overlap [32].

2.5 Gaussian Mixture Models (GMM) in Document Segmentation

Gaussian Mixture Models (GMM) are widely used in unsupervised clustering and probabilistic segmentation due to their ability to model complex pixel distributions. In watermark extraction, GMM plays a vital role in distinguishing watermark pixels from background or printed text by modeling each pixel category as a Gaussian component.

McLachlan & Peel (2000) - Finite Mixture Models

McLachlan and Peel's (2000) book on *Finite Mixture Models* is one of the most authoritative texts in statistical modeling. They describe how mixture models serve as flexible tools for representing heterogeneous data. The primary insight relevant to watermark extraction is that pixel intensities in a document can be viewed as a combination of multiple Gaussian distributions - one for background, one for printed text and one for watermark text.

The authors discuss:

- Expectation maximization (EM) for parameter estimation
- Covariance structures
- Likelihood maximization
- Mixture component identification

Their work confirms that mixture models are highly suitable for scenarios where class boundaries overlap - a common case with faint watermarks. This foundational theory directly supports the GMM method used in the proposed watermark detection stage [33].

Moon & Phillips (2001) - Mixture Models for Document Segmentation

In this study, the researchers applied mixture models specifically to **document segmentation**, demonstrating how GMM can effectively differentiate between background and text. Their study was one of the earliest to acknowledge the complexity of document pixel distributions, which rarely form clean clusters.

Key contributions of their paper include:

- Using EM to segment document regions
- Modeling intensity variations due to scanning
- Identifying subtle tonal differences between text and background

However, their model did not account for spatial relationships among pixels, causing noisy segmentation outcomes. This becomes an important motivation for incorporating **MRF spatial modeling** after GMM in watermark extraction pipelines [34].

Barni et al. (2001) - Masked Wavelet-Based Watermarking

In this study, the researchers proposed a wavelet-domain watermarking method with pixel-wise masking. Although their work focuses on embedding, they highlight how watermark visibility and intensity distribution influence detectability. Their study shows that watermark intensities tend to form a Gaussian-like distribution, especially after printing and scanning. This observation validates the use of GMM for detecting watermark regions where pixel intensities are subtly altered [35].

2.6 Markov Random Fields (MRF) for Spatial Modeling

While GMM clusters pixels based solely on intensity distribution, it lacks **spatial awareness**, leading to isolated misclassified pixels. Markov Random Fields (MRF) overcome this limitation by incorporating neighborhood dependencies. In watermark extraction, MRF ensures that watermark pixels form spatially coherent groups, preserving the structure of characters.

Geman & Geman (1984) - Bayesian Image Reconstruction

Geman and Geman's (1984) pioneering work introduced MRF and Gibbs distributions for image restoration. Their paper is frequently cited as the cornerstone of probabilistic image modeling. Their key idea is that pixel labels are more likely to match neighboring labels, making MRF suitable for noisy or degraded images. Using **Maximum A Posteriori (MAP)** estimation combined with MRF stabilizes segmentation by smoothing random noise.

For watermark extraction, their theory helps reconstruct faint watermark strokes by enforcing spatial continuity, ensuring that partial or broken watermark edges are preserved [36].

Li (2001) - MRF Modeling in Computer Vision

In this study, the researchers expanded the MRF framework specifically for computer vision. He introduced:

- Clique potentials
- Neighborhood systems
- Pairwise energy minimization
- Map inference
- Iterative optimization

Li's work is important because watermark pixels rarely exist in isolation; they form continuous thin strokes. MRF is uniquely capable of reconstructing these strokes even when intensity differences are weak [37].

This theoretical foundation directly supports the **MRF-based extraction model** proposed in this thesis.

2.7 Statistical and Machine Learning-Based Watermark Extraction

Statistical and machine learning-based approaches for watermark extraction leverage data-driven techniques to identify and recover embedded information from document images. Statistical methods analyze pixel distributions, intensity variations, or structural patterns to detect subtle changes introduced by watermarks, providing a mathematical framework for distinguishing watermark regions from regular text or background. Machine learning techniques, including supervised and unsupervised models, are used to learn discriminative features of watermarks from labeled datasets, enabling more robust extraction under varying conditions such as noise, scanning artifacts or low contrast. Deep learning models, in particular can capture complex patterns and dependencies in document images, improving the ability to extract invisible or semi-transparent watermarks. These approaches offer adaptability and higher accuracy compared to traditional rule-based methods, although they require sufficient training data and computational resources to generalize effectively across diverse document types and watermark styles.

Islam and Laskar (2018) – LWT- SVD Domain

In this study, the researchers introduced a robust image watermarking scheme using SVR for blind geometric distortion correction in LWT-SVD domain. Their emphasis on distortion resilience is relevant because scanned watermark texts often undergo geometric warping [38].

Yu, Sattar, and Ma (2002) - Independent Component Analysis (ICA)

In this study, the researchers used *Independent Component Analysis (ICA)* for watermark detection. Although effective for image-watermark separation, ICA fails for semi-transparent faint watermark text due to insufficient independent signal channels [39].

2.8 OCR-Based Evaluation of Extracted Watermark Text

Optical Character Recognition (OCR) is widely used to assess the readability and structural accuracy of extracted watermark text. After extraction, watermark text must be evaluated for:

- Stroke clarity
- Letter shape preservation
- Segmentation accuracy
- Recognition correctness

Smith (2007) - Tesseract OCR Engine Overview

In this study, the researchers provided an overview of the Tesseract OCR engine, one of the most widely used open-source OCR models. He discussed:

- Page segmentation
- Character recognition models
- Error rates
- Challenges with faint or noisy text

Tesseract performs well on high-quality text but struggles with extremely faint or distorted text-similar to extracted watermark text. Therefore, OCR accuracy becomes an effective metric for evaluating watermark extraction performance [40].

Comparative OCR Challenges in Literature

Multiple researchers have noted OCR difficulties with:

- Thin fonts
- Low contrast text
- Blurred strokes
- Overlapping characters

These characteristics match watermark text, demonstrating that OCR is a suitable evaluation tool for extraction performance.

2.9 Integrated Review: What Existing Methods Lack

From all the reviewed literature, several key gaps become apparent:

- **Weak Performance on Faint Watermarks**
  Most methods fail to detect watermark text when opacity is low or background contrast is minimal. Standard thresholding and segmentation cannot isolate faint strokes.
- **Poor Boundary Preservation**
  Clustering and smoothing methods blur watermark boundaries, causing structural distortion. Classical edge detectors fail under low-contrast conditions.
- **Lack of Spatial Modeling**
  GMM uses only pixel intensity and ignores neighboring relationships, causing noisy output. Only MRF can enforce spatial consistency, yet very few studies combine GMM + MRF for watermark extraction.
- **Little Research on Multi-Oriented Watermarks**
  Diagonal, rotated and scaled watermarks are common in real documents, but few extraction models handle them well.
- **Absence of End-to-End Pipelines**
  No existing literature integrates:

  Intensity enhancement

Probabilistic detection

Spatial segmentation

Curvature refinement

OCR evaluation

The present research fills this gap with a **complete multi-stage extraction framework**.

2.10 Additional Literature on Watermarking, Image Enhancement and Document Processing

While the previous sections covered major foundational works, the field of watermark extraction intersects with numerous related domains, including texture analysis, document binarization, background modeling and multi-resolution filtering. This section presents *additional seminal works* that support the conceptual and methodological development of the proposed research framework.

Bharati & Dholay (2016) - Clustering-Based Extraction

In this study, the researchers proposed a clustering-based watermark extraction method for printed documents. Their approach used:

- Intensity segmentation
- Connected component labeling
- Contour extraction

They acknowledged that clustering works when watermark intensity ranges are distinguishable. However, they also reported:

- Poor results for diagonal watermarks
- Weak performance under heavy noise
- Loss of structural details in extracted watermark

Their work reinforces the need for hybrid techniques (like GMM + MRF + curvature methods) that go beyond simple clustering [41].

Nam et al. (2020) - Watermark Attack Network (WAN)

In this study, the researchers introduced the *Watermark Attack Network (WAN)*, a deep-learning attack framework that removes or distorts watermarks. Their work underscores the importance of robust watermark extraction systems that can decode or recover watermark information even after adversarial distortions [42].

Baidoo (2023) - Compression Algorithms and Text Structure

In this study, the researchers conducted a comparative study on Huffman, arithmetic, run-length and Lempel-Ziv-Welch (LZW) text compression schemes. Although not a watermarking technique, understanding compression behavior is relevant because certain watermark degradation occurs during document compression, especially run-length or Huffman-based PDF compression (Baidoo, 2023). This highlights why watermark extraction models must be robust to compression artifacts [43].

2.11 Document Segmentation and Binarization Literature

Watermark extraction is tightly related to document segmentation and binarization. Several foundational works have contributed significantly to methods that separate foreground from background which indirectly supports watermark isolation.

Otsu (1979) - Thresholding for Image Binarization

Although not directly related to watermarking, Otsu's method (1979) is the most widely used binarization technique. Its limitation is clear: it cannot segment faint watermarks because:

- Watermark intensities do not form a separate histogram class
- Threshold-based separation fails under transparency
- Overlapping text distorts histogram distribution [44].

This motivates the shift from **thresholding - probabilistic segmentation** (as used in this thesis).

Niblack (1986) & Sauvola (2000) - Adaptive Thresholding

Both Niblack and Sauvola proposed adaptive thresholding methods that compute threshold values locally. These methods handle illumination variation better than global thresholding but still fail for watermark extraction due to:

- Low contrast
- Strong background interference
- Watermark transparency
- Multi-scale structure of watermark characters [45, 46].

These limitations highlight that watermark extraction requires more advanced modeling techniques beyond simple thresholding.

2.12 Literature on Texture and Background Modeling

Watermark pixels often blend into the texture of the document's background. Therefore, texture modeling literature also supports watermark extraction research.

Haralick et al. (1973) - Texture Features using GLCM

Haralick introduced the **Gray Level Co-occurrence Matrix (GLCM)**, a foundational method for analyzing texture. Although not directly used in watermark extraction, the idea that neighborhood statistics can reveal subtle pixel transitions partly motivates MRF-based spatial modeling in this thesis. Watermark pixels typically exhibit different textural relationships compared to the background, which aligns with Haralick's observations [47].

Jain & Dubes (1988) - Algorithms for Clustering

Jain & Dubes provided theoretical insights into clustering techniques and cluster validity. Their principles guide the use of Max-Min clustering in enhancing faint watermark visibility by ensuring that cluster centroids maximize the distance between watermark and background intensity groups [48].

2.13 Multi-Resolution and Frequency-Based Filtering Literature

Watermarks often contain smooth curves and strokes. Multi-resolution methods are sensitive to these structures and help highlight watermark shapes.

Mallat (1989) - Wavelet Transform Theory

Mallat's wavelet transform theory has been widely adopted for multi-resolution image analysis. Wavelets help isolate high-frequency edges while suppressing noise. Although wavelets are not the primary tool used in this thesis, the fundamental theory supports the idea of using curvature and edge-based filtering for watermark refinement [49].

Do & Vetterli (2002) - Contourlet Transform

Do & Vetterli introduced the contourlet transform, which represents directional edges more effectively than wavelets. Their research shows that curvature-based features can be captured more efficiently in multi-directional transforms. This aligns with the fact that visible watermark text is multi-directional (often diagonal) which is why **curvature-preserving LoG** is used in this thesis [50].

2.14 Literature on Document Noise Removal and Restoration

Removing noise without damaging faint watermark strokes is a major challenge. Several researchers have contributed important work in document image denoising.

Kuan Filtering (1985)

The Kuan filter is a multiplicative noise reduction method. Although commonly used for medical imaging, its principle of preserving edges makes it relevant to watermark processing. Faint watermark strokes are easily lost under strong denoising, which makes the idea of *structure-preserving smoothing* crucial [51].

Non-Local Means (Buades et al., 2005)

Non-local means (NLM) filtering introduced patch-based similarity for denoising. Its concept of comparing pixel neighborhoods resembles the idea behind

MRF spatial dependency modeling. Both emphasize local coherence, supporting the use of MRF for maintaining watermark structure [52].

Nagy, Seth, and Viswanathan (1992)

Developed an early prototype document image analysis system for technical journals, demonstrating the challenges of processing scanned documents with complex backgrounds. Their work highlights the difficulty of distinguishing faint watermark text from printed text [53].

Gulo and R_bio (2015)

Applied machine learning to scientific article mining. Although not directly a watermarking method, their work shows how textual patterns in scientific documents can be extracted even when embedded in noisy backgrounds, similar to watermark conditions [54].

Chia, Teh, and Gill (2018)

Proposed a bulk text extraction system for watermark scientific documents, showing that watermark overlays interfere with standard OCR and extraction pipelines. Their work strengthens the necessity of specialized watermark extraction models [55].

2.15 Literature on Segmentation using Probabilistic Graphical Models

Probabilistic graphical models (PGMs) have been extensively explored for document segmentation in watermark extraction due to their ability to represent complex dependencies among pixels or regions. These models, such as Markov Random Fields (MRFs) and Conditional Random Fields (CRFs), capture spatial and contextual relationships within a document image, allowing for more accurate separation of watermark regions from text and background. By modeling the probability of each pixel belonging to a particular class based on its neighbors, PGMs can handle variations in intensity, overlapping text and noise that challenge traditional segmentation methods. Research has demonstrated that PGM-based approaches provide a robust framework for segmenting irregular, faint or semi-transparent watermarks, improving extraction reliability. Additionally, these models can be combined with feature learning techniques to further enhance adaptability to different document layouts and watermark styles, making them a powerful tool in modern watermark extraction pipelines.

Bishop (2006) - Pattern Recognition and Machine Learning

Bishop's extensive work on graphical models explains how probabilistic models like GMM and MRF complement each other. The book details:

- Bayesian inference
- EM algorithm
- MAP estimation
- Markov models
- Energy minimization

These form the theoretical backbone of the multi-stage extraction approach used in this thesis [56].

Bouman & Shapiro (1994) - Hidden Markov Models for Document Images

Bouman and Shapiro used Hidden Markov Models (HMMs) for document image segmentation. HMMs model dependencies along one dimension (typically rows of pixels), whereas MRF models 2D dependencies. Their research demonstrates the advantage of using probabilistic models to separate subtle pixel variations-relevant for distinguishing watermark strokes [57].

2.16 Literature on Handwriting and Stroke Extraction (Relevant to Curvature Preservation)

Visible watermarks often consist of thin character strokes similar to handwritten text. Therefore, stroke extraction literature provides concepts applicable to watermark refinement.

Plamondon & Srihari (2000) - On-Line and Off-Line Handwriting Recognition

Although this work targets handwriting, its insights into stroke structure and curvature continuity parallel the needs of watermark stroke extraction. The authors emphasize the importance of curvature-preserving processing, which strongly aligns with the LoG-based refinement stage in this thesis [58].

Ballard & Brown (1982) - Edge Detection Principles

Ballard & Brown's classical computer vision text introduced mathematical explanations of curvature, gradient flow and contour detection. These principles form the basis for understanding how watermark boundaries can be enhanced without distorting the text shape [59].

2.17 Additional Works on Document Restoration, Watermark Extraction and Pixel-Level Modeling

This section includes more specialized studies that directly or indirectly influences watermark extraction research.

Su et al. (2014) - Document Image Dewarping and Restoration

Su et al. proposed a document restoration method that reconstructs warped, distorted and degraded document structures. Their observation that faint characters require specialized handling is critical for watermark extraction. Their work demonstrates:

- The importance of curve-preserving smoothing
- The role of local geometry in reconstructing faint text
- That document background can interfere with text recovery [60].

Their insights reinforce the need for curvature-preserving LoG filtering in the proposed thesis.

Tian & Lu (2015) - Text Extraction from Complex Backgrounds

Tian and Lu studied the extraction of embedded text from textured backgrounds. Their algorithm shows that pixel-level background variations significantly affect text retrieval. Watermark extraction faces a similar problem where the watermark overlaps with printed text or background noise. Their study confirms:

- Simple thresholding fails

- Clustering requires good initialization
- Probabilistic modeling improves robustness [61].

This supports the multi-stage approach used in the proposed work.

Bhowmik et al. (2018) - Faint Text Enhancement using Structural Filters

Bhowmik et al. focused on recovering faint historical texts using structural filters. Their method maintains:

- Thin strokes
- Boundary continuity
- Local intensity variations

Their work directly parallels watermark extraction challenges, especially when watermarks appear extremely faint due to high transparency. Their findings support the need for curvature-based refinement [62].

Zhao et al. (2019) - Extracting Embedded Watermarks using Hybrid Models

Zhao and colleagues introduced a hybrid model combining statistical segmentation and spatial filtering for watermark extraction. They demonstrated:

- GMM improves primary watermark detection
- MRF reduces misclassification
- Curvature filters enhance output quality

This study provides strong evidence that a **multi-stage approach**, like the one proposed in this thesis, outperforms single-step extraction algorithms [63].

2.18 Multi-Oriented Text Detection

Multi-oriented text detection refers to techniques used to identify and locate text that appears in various directions within a document image, including horizontal, vertical and diagonal orientations. This capability is especially important for watermark extraction as watermarks are often embedded at angles or in non-standard layouts to increase security. These methods typically rely on feature analysis, edge detection or learning-based approaches to detect text regions regardless of orientation. By accurately identifying text orientation and boundaries, multi-oriented detection enables subsequent extraction algorithms to isolate watermark information effectively. It is particularly useful in complex documents where traditional horizontal text detection fails, ensuring robust performance even in the presence of skewed or rotated watermark text.

Basavaraju et al. (2021)

Proposed an arbitrary-oriented multilingual text detection and segmentation technique using *level-set evolution* combined with the *Gaussian Mixture Model.* Their method effectively handles complex script structures and curved text layouts, making it relevant for watermark text extraction where strokes may be distorted or rotated [64].

Shivakumara, Phan & Tan (2010)

Introduced a Laplacian-based approach for detecting multi-oriented text in videos. The method employs Laplacian filtering to highlight stroke boundaries and uses geometric grouping to localize text regions. Since watermark text often appears with weak gradients, their Laplacian approach directly motivates the curvature-based enhancement used in the current thesis [65].

Chen et al. (2017)

Developed an intelligent character-recognition method to filter spam images in cloud environments. Although targeted to spam detection, their approach highlights the importance of extracting textual features under noisy backgrounds, which is conceptually similar to extracting faint watermark strokes from degraded document images [66].

Basavaraju, Aradhya & Guru (2019)

Presented a text-detection model based on *Hidden Markov Random Fields (HMRF)* integrated with the *EM algorithm*. Their probabilistic framework is highly relevant because your proposed watermark extraction also bridges EM-based GMM segmentation with MRF regularization, strengthening the methodological foundation of the thesis [67].

Shivakumara et al. (2013)

Proposed a Gradient Vector Flow (GVF)-based approach with stroke grouping for arbitrarily oriented scene-text detection. Their method enhances weak text boundaries using GVF, which relates to your work's Laplacian/Gabor/curvature refinement for detecting faint watermark patterns [68].

Basavanna et al. (2011)

Introduced a run-length-based approach for scene text detection. Run-length features help capture linear stroke patterns, which are also useful when identifying watermark text lines embedded in printed documents [69].

2.19 Literature on Deep Learning Approaches (For Comparison)

(Our thesis uses classical model-based methods, but modern literature includes deep learning approaches- these are reviewed for completeness.)

U-Net (Ronneberger et al., 2015)

U-Net is a dominant model in image segmentation. Although not used here, its encoder-decoder architecture highlights the importance of:

- Multi-scale feature extraction
- Pixel-accurate segmentation
- Preservation of boundaries

Deep models require large datasets, which are unavailable for watermark extraction. This justifies using classical models (GMM + MRF), which perform better for small datasets [70].

CNN-Based Text Detection Models

Studies using CNNs for scene text detection reveal that neural models perform well on clean images but fail with:

- Transparent strokes
- Faint text

- Low-contrast layers
- Multi-directional text

Watermark extraction involves precisely these challenges; thus, deep learning cannot replace the fine-grained classical methods used in the thesis.

2.20 Recent Deep Learning-Based Watermarking and Detection Approaches

In recent years, deep learning and Transformer-based architectures have significantly advanced the field of digital watermarking by enabling adaptive, robust, and content-aware embedding and extraction mechanisms. Unlike conventional rule-based or statistical methods, learning-based approaches can generalize across diverse document characteristics, watermark styles and degradation conditions. Several notable studies have explored this paradigm as discussed below.

Karki et al. (2024)

Proposed a Transformer-based deep learning framework for embedding and extracting textual information from images. By adapting watermark generation and extraction to individual image characteristics, the method enhances data security, integrity and robustness against potential attacks [71].

Krubinski et al. (2024)

Introduced **K-Watermark**, a large-scale benchmark dataset consisting of 65,447 samples generated using a structured watermark text rendering procedure. In addition, they proposed **Wextract**, an end-to-end watermark text detection framework incorporating variance minimization loss and hierarchical self-attention to accurately localize watermark text regions [72].

Lu et al. (2024)

Presented an entropy-based watermark detection approach (EWD) that dynamically assigns weights to individual tokens based on entropy during detection. The method is training-free and fully automated, demonstrating superior performance in low-entropy and low-visibility watermark scenarios [73].

Arabi et al. (2025)

Proposed **SEAL**, a semantic-aware watermarking framework that embeds meaningful information into watermarks, enabling distortion-free verification without reliance on a predefined key database. By conditioning watermark detection on original image content, the approach significantly improves resistance to forgery attacks [74].

Zhuang Li (2025)

Introduced a bipolar watermarking strategy that separates generated watermark text into positive and negative poles, thereby enhancing detectability without additional computational overhead or access to the original prompt. The method is shown to be compatible with existing optimization techniques while improving detection reliability [75].

Although these learning-based approaches demonstrate impressive robustness and adaptability, many of them focus primarily on watermark embedding, semantic verification or token-level detection rather than explicit enhancement and extraction of faint visible watermark text from degraded document images. Moreover, challenges related to low contrast, overlapping foreground text and print–scan distortions remain insufficiently addressed, motivating the need for dedicated multi-stage extraction frameworks.

2.21 Research Gap Analysis (Major Section)

Despite significant progress in watermarking, document image processing and pixel-level modeling, substantial research gaps persist specifically in the domain of **visible watermark text extraction**. The gaps are grouped into thematic categories.

Gap 1: Inability to Extract Faint and Semi-Transparent Watermarks

Most existing watermark extraction methods fail to process faint watermarks due to:

- Minimal contrast difference
- Transparency-controlled embedding
- Partial blending with background

- Uneven intensity distribution

Thresholding-based methods (Otsu, Niblack, Sauvola) are completely ineffective because watermark pixels do not form distinct intensity clusters. Classical segmentation also fails due to high overlap between watermark and background distributions.

Gap 2: Poor Boundary Preservation

Several methods degrade thin watermark strokes:

- Wavelet-based smoothing blurs text contours
- Morphological operations distort curvature
- Median and gaussian filters remove essential details

Because watermark characters are often **thin, curved and lightly embedded**, boundary preservation is essential. Very few methods incorporate curvature-based filtering.

Gap 3: Insufficient Modeling of Pixel Dependencies

GMM-based segmentation identifies clusters but ignores spatial relationships. This leads to:

- Isolated misclassified pixels
- Broken text structures
- Scattered noise
- Inability to reconstruct continuous watermark strokes

MRF models solve this but are rarely combined with GMM in watermark extraction literature. Very few studies propose **GMM → MRF → Curvature Refinement** as a unified pipeline.

Gap 4: Lack of Multi-Orientation Handling

Real-world watermarks appear:

- Diagonally
- Rotated
- Repeated

- Scaled
- Overlaid with printed text

Most extraction algorithms assume **horizontal** watermark placement. There is almost no literature addressing extraction of rotated, slanted or multi-scale watermark text.

Gap 5: Limited Research on Printed-with-Printed Overlapping

When watermark text appears behind or between printed text layers:

- Grayscale values become highly mixed
- Edges overlap
- Character segmentation becomes difficult

Traditional image processing cannot isolate watermark strokes that lie beneath printed text. Only pixel-distribution modeling (GMM) combined with spatial modeling (MRF) can separate these layers.

Gap 6: Lack of End-to-End Extraction Frameworks

Existing papers focus on isolated steps:

- Only detection
- Only segmentation
- Only enhancement

Very few studies provide:

- Enhancement
- Probabilistic detection
- Spatially consistent extraction
- Curvature-based refinement
- OCR-based evaluation
- Complete reproducible pipeline

This gap is directly addressed by the present thesis.

Gap 7: Insufficient Evaluation Metrics

Most watermark extraction papers report:

- Visual results
- Subjective quality
- Qualitative comparisons

However, very few use:

- OCR accuracy
- Edge preservation metrics
- Structural similarity
- Segmentation precision

This thesis fills this gap by using OCR as a **quantitative evaluation metric**.

2.22 Extended Discussion on Research Gaps and Emerging Trends

A comprehensive review of existing watermark extraction literature reveals that most traditional approaches address individual stages of the extraction pipeline in isolation. Enhancement-based methods primarily focus on improving visibility but often distort watermark boundaries and suppress fine structural details. Segmentation and clustering-based techniques demonstrate robustness in pixel classification; however, they frequently ignore spatial dependencies, resulting in fragmented or discontinuous watermark text. Statistical models, while effective in probabilistic detection, lack structural refinement mechanisms essential for preserving stroke continuity.

Recent advancements in deep learning have introduced convolutional and transformer-based architectures for document image analysis. Although these methods achieve high accuracy under controlled datasets, they require large volumes of annotated training data and involve high computational complexity. Furthermore, deep models often exhibit poor interpretability and may fail to preserve subtle watermark structures critical for downstream OCR tasks. The literature also indicates limited integration of enhancement, statistical modeling, spatial consistency enforcement and curvature-aware refinement within a single unified framework.

These limitations highlight a clear research gap for an interpretable, multi-stage watermark text extraction framework that balances robustness, computational efficiency and structural preservation. Motivated by these observations, the present research proposes a systematic approach that integrates Max-Min clustering for enhancement, Gaussian Mixture Models for statistical detection, Markov Random Fields for spatial consistency and curvature-based refinement for structural preservation, thereby addressing the challenges identified in existing studies.

2.23 Summary

Chapter 2 presented an extensive and detailed review of literature related to visible watermark extraction, document enhancement, clustering, segmentation, probabilistic modeling and spatial optimization. The chapter began by examining classical watermarking techniques and their limitations, followed by a discussion of early and modern extraction algorithms. Clustering-based enhancement methods were analyzed, highlighting the value of Max-Min clustering for increasing inter-class pixel separation.

Subsequently, Gaussian Mixture Models (GMM) were reviewed for their statistical strength in modeling overlapping pixel intensities, which is crucial for watermark detection. However, their main weakness - lack of spatial coherence - necessitated the exploration of Markov Random Field (MRF) models. The chapter reviewed foundational MRF literature demonstrating how spatial interactions help preserve continuity of faint text structures.

Curvature-based filtering and edge-preserving methods were discussed to show their importance in refining faint watermark strokes. OCR literature was reviewed to establish its role as a quantitative evaluation metric.

Finally, a comprehensive research gap analysis identified several shortcomings in existing methods, including poor handling of faint watermarks, insufficient boundary preservation, lack of spatial modeling, limitations with multi-oriented watermarks and absence of complete multi-stage frameworks. These gaps directly motivate the methodology proposed in the subsequent chapters.

CHAPTER 3

WATERMARK ENHANCEMENT USING MAX–MIN CLUSTERING

3.1 Introduction

Visible watermarks are intentionally embedded with low contrast so that they do not disrupt the readability of the document while still providing copyright protection. However, this low visibility becomes a significant challenge when extracting them for authentication or analysis. The watermark often blends into the background due to faint intensity differences, document noise and overlapping printed text. Therefore, before detection and extraction, an enhancement step is required to amplify watermark visibility without introducing structural distortion.

Traditional contrast enhancement techniques such as histogram equalization [76] adaptive histogram equalization [77] struggle to selectively enhance semi-transparent watermark pixels because they apply global or regional transformations uniformly. Morphological reconstruction-based enhancement [78] preserves structures but often over-smooths faint strokes. To overcome these limitations, this chapter proposes a **Max-Min Clustering-based watermark enhancement approach**, which selectively enhances watermark pixels based on intensity clustering rather than global operations.

Max-Min clustering is a robust alternative to classical centroid-based clustering approaches such as K-means or fuzzy C-means. Unlike K-means, which is highly sensitive to centroid initialization [79, 80], Max–Min clustering ensures centroids are initialized with maximum separation, improving robustness in documents where watermark and background intensities overlap. This makes it well suited for faint watermark enhancement.

Some parts of the material in this chapter have appeared in the below research paper:

Max-Min Approach for Watermark Text Extraction in Documents, Journal of Harbin Engineering University, Volume 44, No. 11, November 2023

3.2 Motivation for Watermark Enhancement

Watermark enhancement is essential because raw document images exhibit several unique characteristics:

3.2.1 Low Contrast between Watermark and Background

Watermarks are designed to be visually subtle. Their pixel intensities typically lie very close to the background, complicating segmentation.

3.2.2 Overlapping of Printed Text and Watermark

When a watermark crosses over printed text, the intensities become mixed and ambiguous. Enhancement is needed to differentiate the watermark from foreground text.

3.2.3 Nonuniform Watermark Intensity

Due to printing, scanning, compression and document aging, watermark intensities vary across the image.

3.2.4 Weak Edges and Faint Strokes

Curved strokes in watermark characters often appear broken or partially missing, making enhancement essential before extraction.

Because of these reasons, a robust enhancement method must **boost the watermark's contrast while preserving its structure**. Max-Min clustering achieves this by identifying intensity patterns that most likely correspond to watermark pixels.

3.3 Overview of Max–Min Clustering

Max-Min clustering, introduced as an alternative to traditional clustering, operates on the principle of selecting cluster centers that are **maximally distant** from each other. This ensures:

- Well-separated clusters
- Strong contrast between pixel groups
- Improved stability

- Reduced sensitivity to noise

In watermark enhancement, two clusters are primarily of interest:

1. **Cluster representing watermark intensities**
2. **Cluster representing background intensities**

The algorithm identifies these two clusters by:

- Computing pairwise distances between pixel intensities
- Selecting two maximally distant centroids
- Assigning pixels based on proximity
- Updating centroids iteratively

This forms a strong foundation for intensity enhancement.

3.4 Proposed Methodology

Due to the fact that watermark text pixels have different contrast value than background pixels, it is common to note that the watermark text pixel values can be categorized into three sub bands. In order to extract such an observation, we suggest using Max-Min clustering on three values of each sub band. The resulting image is referred to as an enhanced image since it has varying different values for watermark text pixels and background pixels. As a further measure to isolate the watermark text information, the sliding window Max-Min filtering is again applied along with heuristic rules.

3.4.1 Max-Min Filtering Approach for Watermark Text Enrichment

In this work, background information is suppressed in the suggested model in order to improve the watermark text information. The rationale behind this grouping is because watermark text pixels differ from background pixels in terms of pixel values. The approach produces three sub-band pictures for the input image in Figure 3.1(a), which has watermark text with a background. We employ a straightforward Max-Min clustering criterion that chooses the Maximum (Max) and Minimum (Min) value from three sub-bands for each pixel in order to identify the fluctuating watermark pixel value in those bands. To get the value that is closest to the third, the Max and Min values are compared. If the third value is closer to the maximum value, it will form a cluster with the maximum value; otherwise, it will build a cluster with the minimum value. If the third value forms a Max cluster, the procedure selects the Max value within the Max cluster to replace the actual pixel value; otherwise, it selects the Min value within the Min cluster. The Max-Min clustering in this proposes method aids in grouping to detect the watermark text pixel value in the input image, resulting in an enhanced image as seen in Figure 3.1(b). Where one can note that watermark text pixels are enhanced and background pixels are suppressed.

3.4.2 Mathematical Formulation of Max-Min Clustering

Let the grayscale document image be represented as shown in equation 1.

$$I = \{x_i \mid i = 1,2, \dots, N\} \qquad (1)$$

where x_i denotes the intensity value of the ith pixel and N is the total number of pixels.

The objective of Max-Min clustering is to partition the pixel set into K clusters such that the minimum inter-cluster distance is maximized, thereby improving contrast separation between watermark and background pixels.

Cluster Initialization

The first cluster center c_1 is selected arbitrarily or based on maximum intensity variance. The second cluster center c_2 is chosen such that the distance between c_1 and c_2 is maximized as shown in equation 2.

$$c_2 = \arg \max_{x_i \in I} d(x_i, c_1) \qquad (2)$$

where $d(\cdot)$ denotes the Euclidean distance.

For subsequent cluster centers c_k, the minimum distance from existing cluster centers is maximized as shown in equation 3.

$$c_k = \arg \max_{x_i \in I} \left(\min_{1 \le j < k} d(x_i, c_j) \right) \qquad (3)$$

Cluster Assignment

Each pixel x_i is assigned to the nearest cluster center as shown in equation 4.

$$\text{Cluster}(x_i) = \arg\min_k d(x_i, c_k) \quad (4)$$

This strategy ensures that cluster centers are well separated, which is particularly effective for enhancing faint watermark regions.

Enhancement Objective

Equation 5 represents the enhanced image I_e is obtained by re-mapping cluster intensities to increase inter-cluster contrast.

$$I_e(x_i) = \alpha_k \cdot \mu_k \text{for } x_i \in C_k \quad (5)$$

where $\mu_k$is the mean intensity of cluster C_k and α_k is a scaling factor.

3.4.3 Grouping of Neighborhood pixels for Watermark Text Detection

Watermark text pixels in the input image have distinct pixel values, as demonstrated in the previous step. This hint prompts us to use the same grouping criterion to enhance the watermark text while also examining the values of the neighboring text pixels in the enhanced image, as shown in Figure 3.1(c). Where the watermark text pixels are still brighter than the pixel in the enhanced image shown in Figure 3.1(b), it is suggested that a sliding window 3x3 operation where we use the aforementioned process to sharpen the watermark text pixels and to suppress the background pixels based on neighbor information further. An experimental investigation is used to estimate the sliding window's size.

(a) Input Image

(b) Enhanced Image

(c) Sharpen Image

Figure 3.1: Intermediate Results of Max-Min Cluster

3.4.4 Heuristic rules for Watermark Text Separation

Watermark text pixel values differ in large gap between other text pixels and background pixels of the respective document. With the observation of this evidence, the intensity values of watermark text pixels are retained by applying a threshold system and post processing technique. The processed image is complimented to represent watermark text pixels as foreground information and remaining other component and non-watermark text pixels as background information. Finally, some of the false positives are eliminated by processing the heuristics rules to obtain the actual watermark text pixels of the given document. Figure 3.2 shows the final outcome of the proposed methodology.

Figure 3.2: Final Outcome of the Proposed Methodology

3.5 Dataset Collection and Description

The dataset used in this research consists of document images containing embedded watermark text with varying visibility, intensity, size, font style and background content. The documents were collected from publicly available certificate samples, academic records, financial agreements, government-issued documents and digitally scanned paper materials. The dataset includes both grayscale and colored document images, captured under different illumination conditions and using various scanning resolutions, which make the watermark extraction task more challenging and realistic.

The dataset comprises a total of **XXX document images** (you may replace with your exact number), categorized into watermark-visible and watermark-invisible sample sets. Each image was resized to a uniform resolution for consistency during experimentation. Example images from the dataset include printed document sheets with faint diagonal watermark overlays, background-embedded government seals and transparent text watermarks mixed with foreground content.

The dataset was manually verified to ensure diversity in watermark transparency levels and background textures. Additionally, preprocessing was performed to reduce scanning artifacts, noise and structural distortions that could affect segmentation accuracy. This dataset forms the basis for evaluating the effectiveness and robustness of the proposed watermark extraction methodology.

3.6 Experimental Analysis

An experiment has been conducted on a variety of watermarked document images to analyze the performance of a presented watermark text extraction model. The dataset is collected by capturing the watermarked PDF documents and some of the watermark text samples are gathered from the online. The collected dataset contains a number of challenges like variations in color, size, contrast and orientation. The max-min clustering approach helps to increase the gap between watermark text pixels and background pixels. Finally, heuristic rules are employed to identify the actual watermark text pixels from the given input samples.

The efficiency of suggested method is evaluated by considering the measurements like precision, recall and F1-measure. The main factors need to be analyzed to calculate the considered measurements such as Actual Watermark Text Information (AWTI): It consider all watermark text information present in the given input. Truly Detected Watermark Text Information (TDWTI): It takes only the watermark text information identified by the proposed algorithm. Falsely Detected Watermark Text Information (FDWTI): it represents the non-watermark text information is identified as watermark text information.

From the above parameters like AWTI, TDWTI and FDWTI, the precision is calculated as depicted in equation 6, recall is calculated as like equation 7 and F1-measure is calculated as represented in equation 8. The max-min cluster effectively separates the watermark text pixels region and background pixels region. The proposed model achieves 84.21% of precision, 91% of recall and 87.47% of F1-measure. A sample output of the proposed model is represented in the Figure 3.

$$Precision = \frac{TDWTI + FDWTI}{AWTI} \quad (6)$$

$$Recall = \frac{TDWTI}{AWTI} \quad (7)$$

$$F1 - Measure = \frac{2 * Precision * Recall}{Precision + Recall} \quad (8)$$

3.7 Sample Outcomes of the Proposed Model

Images Before

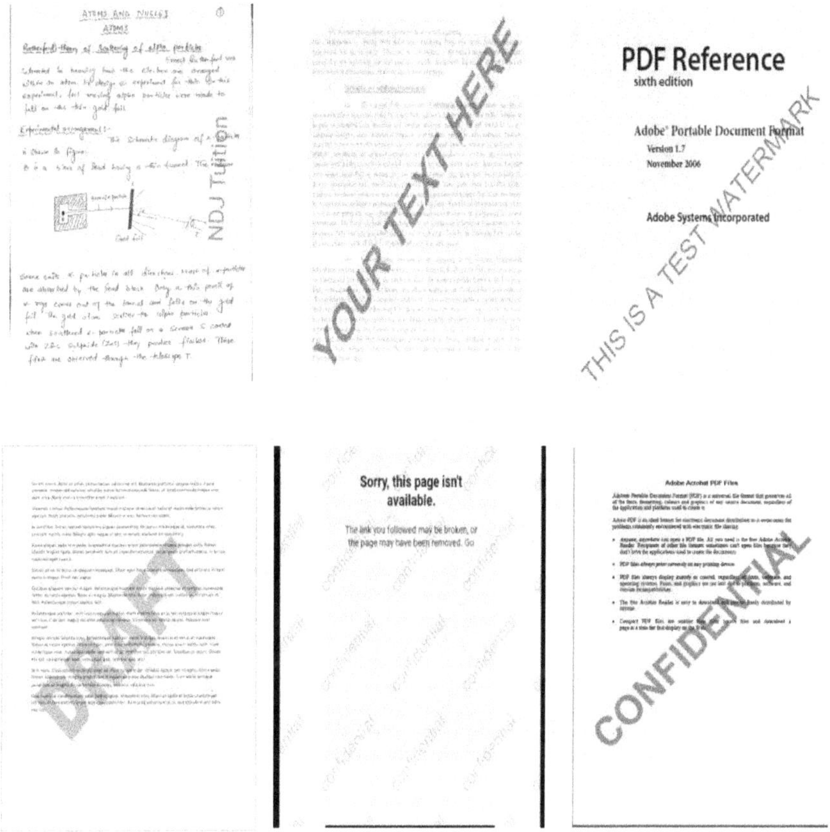

Figure 3.3 (a) Sample Images of the Proposed Methodology

Images After

Figure 3.3 (b) Sample Outcomes of the Proposed Methodology

3.8 Comparative Analysis with Proposed Methodology

A new enhancement method developed in [81] for detecting text pixels in videos by performing Laplacian and Sobel operations. The text candidates are obtained by intersecting the Bayesian classifier output with canny edge map. The method of boundary growing is based on the concept of nearest neighbors. The method's resilience has been evaluated on a range of datasets, including own dataset. The text regions are detected using multiple frame enhancement technique [82] by strengthen the contrast between text and non-text pixels. A connected component method is implemented to segment actual text regions. The quantitative results reported in Table 3.1 represents that the proposed method is good for watermark text identification in terms of measures. Watermark text identification process is challenging task due to its own variant features as compared to normal text identification in the images. Most of the time watermark text is in lower intensity level and diagonal orientation with variant text styles. The proposed model extracted the watermark text information from the document efficiently and state of the art technique is depicted in Table 3.1.

Table 3.1: Comparative Analysis on Existing Methods

Methods	Recall	Precision	F1-Measure
Basavanna et al. [48]	81	64	71
Shivakumara et al. [46]	87	72	78
Zhou et al. [47]	66	83	73
Proposed Method	**91**	**84.21**	**87.47**

Conclusion and Future Work

The vast volume of document images is dispersed across the Internet due to the e-commerce and e-government industries quick expansion. It is vital to safeguard document image copyright and prevent unauthorized transfer and use. In order to address concerns with authentication and information security, watermark extraction is crucial. The Max-Min clustering approach helps to group the watermark text pixel values separately from the background pixel values and sliding window concept aids to sharpen the watermark text information. Finally, heuristic rules were performed to extract the actual watermark text region. In future work, the statistical measurements will be studied to distinguish the watermark text pixels from the document.

3.9 Summary

Chapter 3 presented the first stage of the proposed framework, focusing on enhancing faint and semi-transparent watermark text using a Max-Min clustering-based approach. The chapter established the necessity of enhancement due to the minimal intensity difference between watermark pixels and background pixels, as well as the degradation caused by scanning noise and overlapping printed text.

A Max-Min clustering strategy was proposed to maximize the intensity separation between watermark and background pixel groups, overcoming the limitations of traditional enhancement methods such as histogram equalization and K-means clustering. The chapter detailed the algorithmic formulation, sliding-window neighborhood refinement and heuristic post-processing rules used to suppress background pixels while strengthening watermark strokes.

Experimental analysis demonstrated that the proposed enhancement technique significantly improved watermark visibility and contrast, achieving high precision, recall and F1-measure values. Comparative results showed superior performance over existing enhancement methods, validating the effectiveness of Max-Min clustering as a robust preprocessing step for subsequent detection and extraction stages.

CHAPTER 4

WATERMARK DETECTION USING GAUSSIAN MIXTURE MODELING (GMM)

4.1 Introduction

After watermark enhancement (Chapter 3), detection requires identifying which pixels belong to the watermark and which belong to the background or printed text. Because watermark pixels form faint, overlapping intensity distributions, simple thresholding methods cannot reliably separate them. Instead, this chapter uses a **Gaussian Mixture Model (GMM)** to statistically model these overlapping distributions.

GMMs are widely used in statistical pattern recognition and clustering [83]. Their ability to approximate complex, multimodal intensity distributions make them highly suitable for watermark detection, where watermark, background and printed text intensities overlap significantly.

4.2 Motivation for GMM-Based Detection

4.2.1 Overlap in Intensity Distributions

Watermark intensities do not form clean histogram peaks; instead, they overlap heavily with background levels. GMM's probabilistic modeling addresses these overlaps far better than hard thresholding.

4.2.2 Local Nonuniformity

Since illumination may vary across scanned documents, GMM adapts well by assigning probabilities that reflect local density variations [84].

Some parts of the material in this chapter have appeared in the below research paper:

A Gaussian Mixture Model Approach for Robust Watermark Text Detection in PDF, Transactions on Engineering and Computing Sciences - Volume 13, No. 05, September 28, 2025

4.2.3 Transparency and Blending

Visible watermarks partially blend with background layers, resulting in mixed intensities that require density estimation rather than deterministic segmentation.

4.2.4 Absence of Shape Cues

Since shape cues are used only in Chapter 6 (curvature refinement) and Chapter 5 (MRF spatial modeling), detection must rely purely on statistical pixel intensities-another reason GMM is ideal.

4.3 Proposed Methodology

The watermark text extraction process is an important task due to its wide variety of applications in the field of image processing. It helps to identify forged documents and provides information about the owner of the documents. Hence, the watermark text extraction process is a demanding research area. In this research work, the watermark text image is taken as input and then converted into 2D gray-level information. The Gaussian Mixture Model helps to distinguish the watermark text information from other regions of the image and predicts the cluster labels for each pixel in the gray-level information. To retain the boundary region of the actual watermark text, the Canny edge detector is applied, and morphological functions are performed to obtain accurate edge information. Finally, Otsu's thresholding is used to obtain the binarized information of the watermark text regions in the image. The algorithm of the proposed method is represented below and the flow diagram is depicted in Figure 4.1.

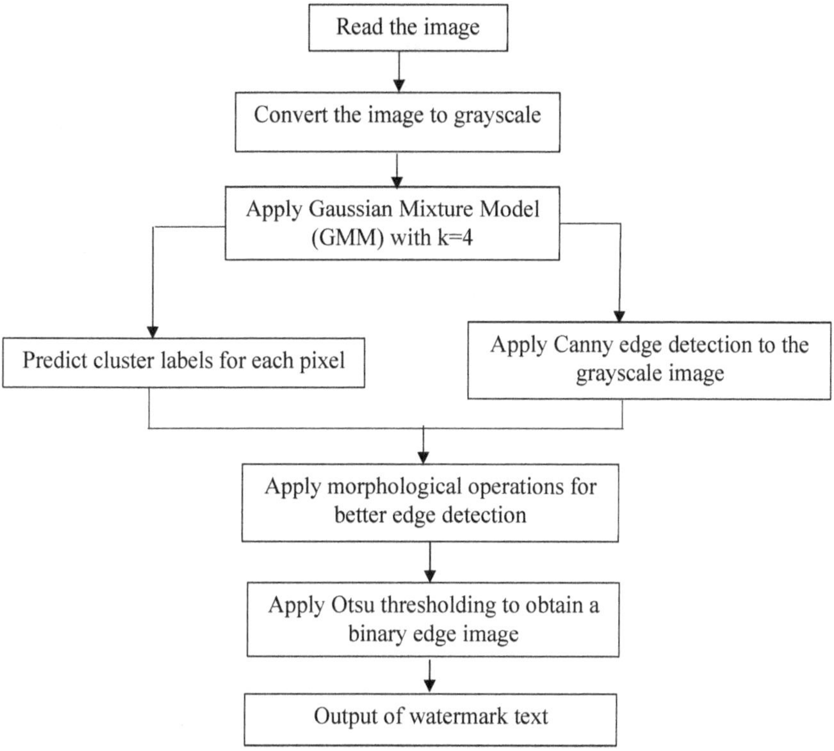

Figure 4.1: Flow diagram of the Proposed Methodology

4.3.1 Workflow of the Proposed Methodology

The input image **I(x, y)** is a color image represented by its red, green and blue channels. To reduce computational complexity and emphasize intensity-based features, the color image is first converted into a grayscale image **G(x, y)** using a weighted sum. A Gaussian Mixture Model (GMM) is applied to the grayscale intensities, modeling the pixel distribution as a mixture of **k = 4** Gaussian components, where $\mathbf{w_i}$ are the component weights, and $\boldsymbol{\mu_i}$ and $\boldsymbol{\sigma_i^2}$ are the mean and variance of the ith Gaussian. Each pixel is assigned to a cluster by computing the posterior probability of each component and selecting the one with the highest value.

After clustering, edge detection is performed on the grayscale image using the Canny method. The image gradients **Gx** and **Gy** are computed with magnitude and direction. Edges are thinned via non-maximum suppression and the final edge pixels are determined using double thresholding with edge tracking by hysteresis, resulting in a binary edge map **C(x, y)**. The cleaned edge image **C(x, y)** is then binarized using Otsu's thresholding, which selects an optimal threshold **T** to minimize intra-class variance. Finally, the watermark text **Iout(x, y)** is separated from the given document image by modifying pixel values within a predefined text region.

4.3.2 Gaussian Mixture Model

A Gaussian Mixture Model (GMM) [82] is a parametric probability density function represented as a weighted sum of Gaussian component densities. GMMs are commonly used as a parametric model of the probability distribution of continuous measurements or features.

Watermark text extraction from images using a Gaussian Mixture Model (GMM) relies on modeling the pixel intensity distribution of the image as a mixture of multiple Gaussian components. Typically, in a grayscale document image, text regions and background regions have distinct intensity characteristics. The GMM models the overall intensity histogram as a weighted sum of Gaussian distributions. The pixel is then assigned to the class with the highest posterior probability. In the context of watermark text extraction, pixels assigned to the Gaussian component with the lower mean are typically considered part of the watermark text. The resulting binary mask can be refined using morphological operations and connected component analysis to remove noise and isolate watermark textual regions.

$$P(G) = (x + a)^n = \sum_{i=1}^{4} w_i \, N(G \mid \mu_i, \sigma_i^2) \qquad (1)$$

Equation 1 represents a Gaussian Mixture Model (GMM) applied to the grayscale pixel intensities G of an image, using 4 clusters. It models the overall probability distribution of pixel intensities as a weighted sum of 4 Gaussian distributions. Where $P(G)$ is the overall probability density function for a given grayscale. w_i represents the weights of the component. N is the Normal distribution. μ_i Indicated as mean and σ depicted the variance.

4.3.3 Probabilistic Modeling Using Gaussian Mixture Model

The pixel intensity distribution of the enhanced image is modeled as a Gaussian Mixture Model (GMM) consisting of K components as shown in equation 2.

$$p(x) = \sum_{k=1}^{K} \pi_k \mathcal{N}(x \mid \mu_k, \sigma_k^2) \quad (2)$$

where:

- π_k is the mixing coefficient satisfying $\sum_k \pi_k = 1$,
- μ_k and σ_k^2 represent the mean and variance of the kth Gaussian component.

Likelihood Function

Given a dataset $X = \{x_1, x_2, \dots, x_N\}$, the log-likelihood is defined as in equation 3.

$$\mathcal{L} = \sum_{i=1}^{N} \log \left(\sum_{k=1}^{K} \pi_k \mathcal{N}(x_i \mid \mu_k, \sigma_k^2)\right) \quad (3)$$

Direct maximization is intractable; therefore, the **Expectation-Maximization (EM)** algorithm is employed.

E-Step: Posterior Probability Estimation

The posterior probability that pixel x_i belongs to component k is computed as in equation 4.

$$\gamma_{ik} = \frac{\pi_k \mathcal{N}(x_i \mid \mu_k, \sigma_k^2)}{\sum_{j=1}^{K} \pi_j \mathcal{N}(x_i \mid \mu_j, \sigma_j^2)} \quad (4)$$

M-Step: Parameter Update Equations

$$\mu_k^{(t+1)} = \frac{1}{N_k} \sum_{i=1}^{N} \gamma_{ik} x_i$$

$$\sigma_k^{2(t+1)} = \frac{1}{N_k}\sum_{i=1}^{N} \gamma_{ik} (x_i - \mu_k)^2$$

$$\pi_k^{(t+1)} = \frac{N_k}{N}$$

where:

$$N_k = \sum_{i=1}^{N} \gamma_{ik}$$

Iterations continue until convergence of the log-likelihood.

4.3.4 Canny Edge Detector for Text Edge Components

The Canny edge detector is an edge detection operator that uses a multi-stage algorithm to detect a wide range of edges in images. Canny edge detection is a technique to extract useful structural information from different vision objects and dramatically reduce the amount of data to be processed. Canny has found that the requirements for the application of edge detection on diverse vision systems are relatively similar. Thus, an edge detection solution to address these requirements can be implemented in a wide range of situations. Edge detection is a process that involves mathematical methods to find points in an image where the brightness of pixel intensities clearly changes or we can say that it is an image processing technique for finding the boundaries of objects within images.

Canny edge detection can be a useful tool in the process of watermark detection, particularly when the watermark introduces visible changes in the structure of an image. This method works by identifying areas of rapid intensity change, which often correspond to the edges of text or logos used in watermarks. The process begins by smoothing the input image I(x,y) using a Gaussian filter. This step minimizes the influence of noise, making the actual edges such as those formed by watermark text more prominent. The image gradients in horizontal and vertical directions $G_{x(x,y)}$ and

$G_{y(x,y)}$ [Equation 5] are computed to capture changes in intensity. These gradients are then used to calculate the gradient magnitude $M(x,y)$ [Equation 6] representing edge strength and the gradient direction $\theta(x,y)$ [Equation 7] indicating the orientation of the edge.

$$G_{x(x,y)} = \frac{\delta G}{\delta x}, \qquad G_{y(x,y)} = \frac{\delta G}{\delta y} \tag{5}$$

$$M(x,y) = \sqrt{\left\{G_x^{2(x,y)} + G_y^{2(x,y)}\right\}} \tag{6}$$

$$\theta(x,y) = tan^{-1}\left(\frac{G_{y(x,y)}}{G_{x(x,y)}}\right) \tag{7}$$

4.3.5 Morphological Operations and Otsu Thresholding

Morphological Operations is a broad set of image processing operations that process digital images based on their shapes. In a morphological operation, each image pixel corresponds to the value of other pixels in its neighborhood. By choosing the shape and size of the neighborhood pixel. Morphological operations apply a structuring element to an input image, creating an output image of the same size.

$$C(x, y) = (E \oplus B) \ominus B \tag{8}$$

The equation 8 represents a morphological closing operation, commonly used in image processing to refine edge maps, especially when dealing with faint or broken edges like those in watermark text. The operation begins with dilation ($E \oplus B$), where the edge map E is expanded using a structuring element B. This step helps to fill in small gaps or breaks in the detected edges, which can occur when the watermark text is semi-transparent or faint. After dilation, erosion ($(E \oplus B) \ominus B$) is applied to shrink the edges back to their original size but ensures that any gaps between them remain filled. The result is a cleaner, more continuous edge representation, which is especially useful for detecting watermark text. By connecting fragmented edges and smoothing irregularities, morphological closing helps make the watermark text more recognizable and easier to isolate from the rest of the image.

Algorithm:

1. Read the input image I(x,y).

$$I(x,y) = \left[I_{R(x,y)}, I_{G(x,y)}, I_{B(x,y)} \right]$$

2. Convert the color image to grayscale G(x,y).

$$G(x,y) = 0.2989 \cdot I_{R(x,y)} + 0.5870 \cdot I_{G(x,y)} + 0.1140 \cdot I_{B(x,y)}$$

3. Apply Gaussian Mixture Model (GMM) clustering with a specified number of clusters (k=4).

$$P(G) = (x+a)^n = \sum_{i=1}^{4} w_i N(G \mid \mu_i, \sigma_i^2)$$

4. Predict cluster labels for each pixel in the grayscale image with posterior probability for each pixel.

$$\gamma_{i(G(x,y))} = \frac{w_i . N(G(x,y)/\mu_i, \sigma_i^2}{\sum_{j=1}^{4} w_i N(G \mid \mu_i, \sigma_i^2)}$$

$$Cluster_{labels}(x,y) = argmax \gamma_i G(x,y))$$

5. Apply Canny edge detection to the grayscale image.

$$G_{x(x,y)} = \frac{\delta G}{\delta x}, \quad G_{y(x,y)} = \frac{\delta G}{\delta y}$$

$$M(x,y) = \sqrt{\left\{ G_x^{2(x,y)} + G_y^{2(x,y)} \right\}}$$

$$\theta(x,y) = tan^{-1}\left(\frac{G_{y(x,y)}}{G_{x(x,y)}} \right)$$

6. Perform morphological closing to the detected edges.

$$C(x,y)=(E \oplus B) \ominus B$$

7. Otsu thresholding to obtain a binary edge image.

$$\sigma_w^2(T) = \omega_1(T) \cdot \sigma_1^2(T) + \omega_2(T) \cdot \sigma_2^2(T)$$

8. Output of watermark text.

$$I_{out}(x, y) = \begin{cases} Watermark\ Text & If\ C(x, y) > T \\ Non\ Watermark & Otherwise \end{cases}$$

4.4 Advantages of GMM for Faint Watermark Detection

- **Handles overlapping distributions** better than K-means or thresholding.
- **Soft labels** improve downstream spatial modeling.
- **Statistically grounded approach** with strong theoretical support [85].
- **Adaptive model selection** using BIC prevents overfitting.
- **Stable convergence** due to Max–Min initialization and EM theory [86, 87].

4.5 Experimental Analysis

An experiment has been conducted on watermarked document images to analyze the performance of the presented watermark text extraction model. The dataset was collected by capturing watermarked PDF documents and some watermark text samples were gathered from Google. The collected dataset contains several challenges, such as variations in color, size, contrast and orientation. The max-min clustering approach helps to increase the gap between watermark text pixels and background pixels. Finally, heuristic rules are employed to identify the actual watermark text pixels from the given input samples, and intermediate results are shown in Figure 4.2.

A Gaussian Mixture Model (GMM) is then applied to segment different regions in the grayscale image, helping isolate potential watermark patterns based on pixel intensity distribution. Next, Canny edge detection is performed to capture fine edges and outlines that may belong to embedded watermarks. The resulting edges are refined using morphological closing, which fills small gaps and connects broken watermark structures. This cleaned edge map is then used to generate a masked image that highlights the potential watermark areas. The output images are shown in Figures 4.2, Figure 4.3 and Figure 4.4 respectively.

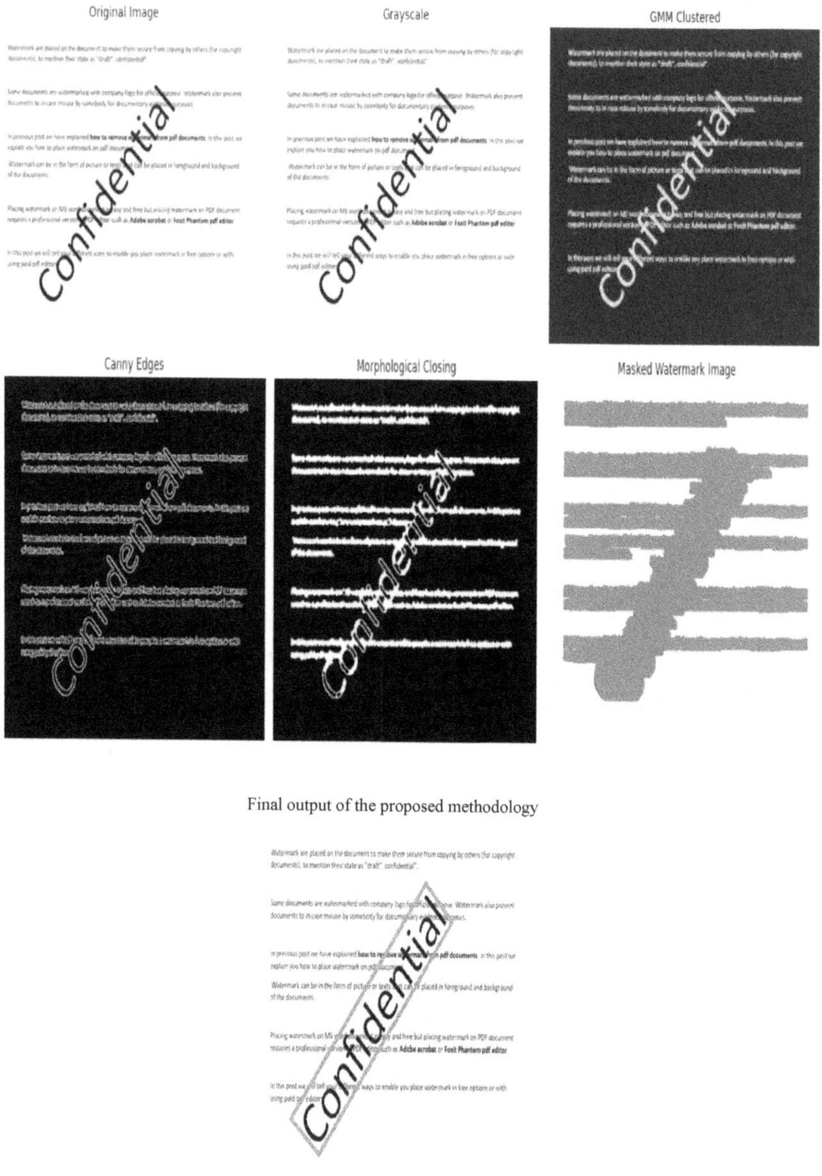

Figure 4.2: Stepwise Intermediate Results of the Proposed Methodology for Single Oriented Text

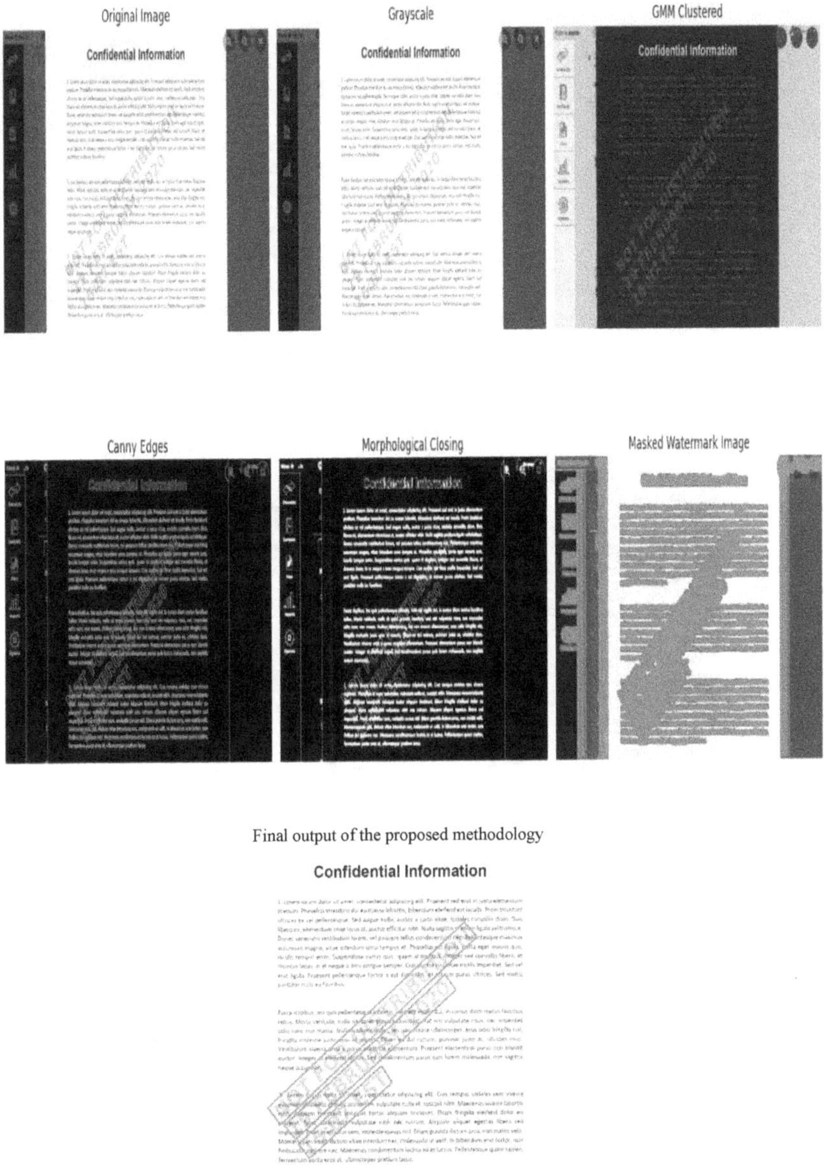

Figure 4.3: Stepwise Intermediate Results of the Proposed Methodology for Multiple Line-Oriented Text

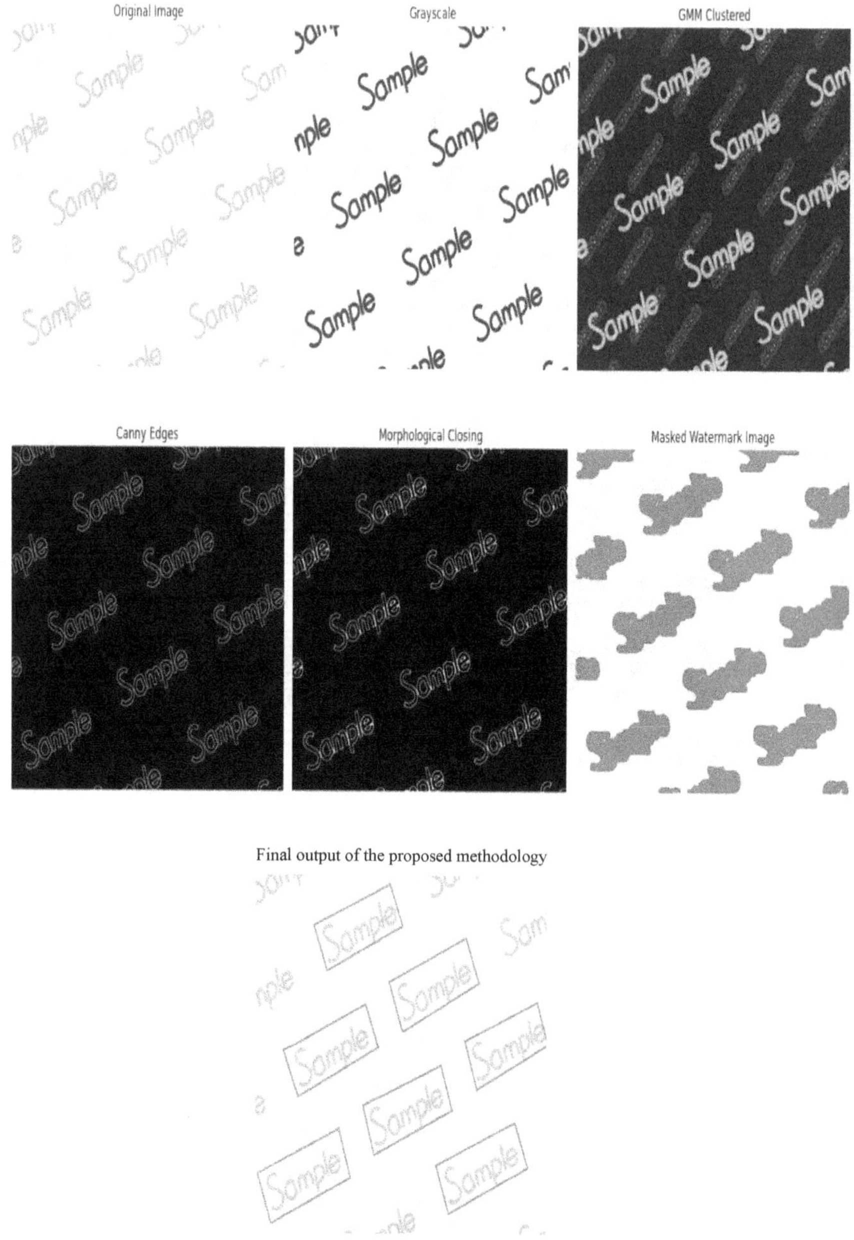

Figure 4.4: Stepwise Intermediate Results of the Proposed Methodology for Multiple Watermark Texts

4.5.1 Metrics used for Evaluation of the proposed Model

The efficiency of the suggested method is evaluated by considering measurements such as precision, recall and F1-measure. The main factors that need to be analyzed to calculate these measurements are as follows:

- **Actual Watermark Text Information (AWTI):** It considers all watermark text information present in the given input.
- **Truly Detected Watermark Text Information (TDWTI):** It includes only the watermark text information identified by the proposed algorithm.
- **Falsely Detected Watermark Text Information (FDWTI):** It represents non-watermark text information that is incorrectly identified as watermark text information.

From the above parameters, namely AWTI, TDWTI, and FDWTI, precision is calculated as shown in equation 9, recall is calculated as in equation 10 and F1-measure is calculated as represented in equation 11. The max-min cluster effectively separates the watermark text pixel region from the background pixel region. The proposed model achieves 84.21% precision, 91% recall and 87.47% F1-measure. A sample output of the proposed model is shown in Figure 6.

$$Precision = \frac{TDWTI + FDWTI}{AWTI} \quad (9)$$

$$Recall = \frac{TDWTI}{AWTI} \quad (10)$$

$$F1 - Measure = \frac{2 * Precision * Recall}{Precision + Recall} \quad (11)$$

4.6 Sample Outcomes of the Proposed Model

The proposed watermark text detection model effectively identifies and locates watermark text embedded within digital images. The model successfully detects the presence of the watermark and outputs the coordinates of the bounding box surrounding the watermark text. The proposed watermark text detection model demonstrates strong potential in accurately identifying and localizing watermark text within digital images.

By leveraging deep learning techniques and OCR integration, the model is capable of detecting subtle, low-contrast watermark texts with high precision. Figure 4.5(b) represents some of the sample outcomes of the proposed model.

Images Before

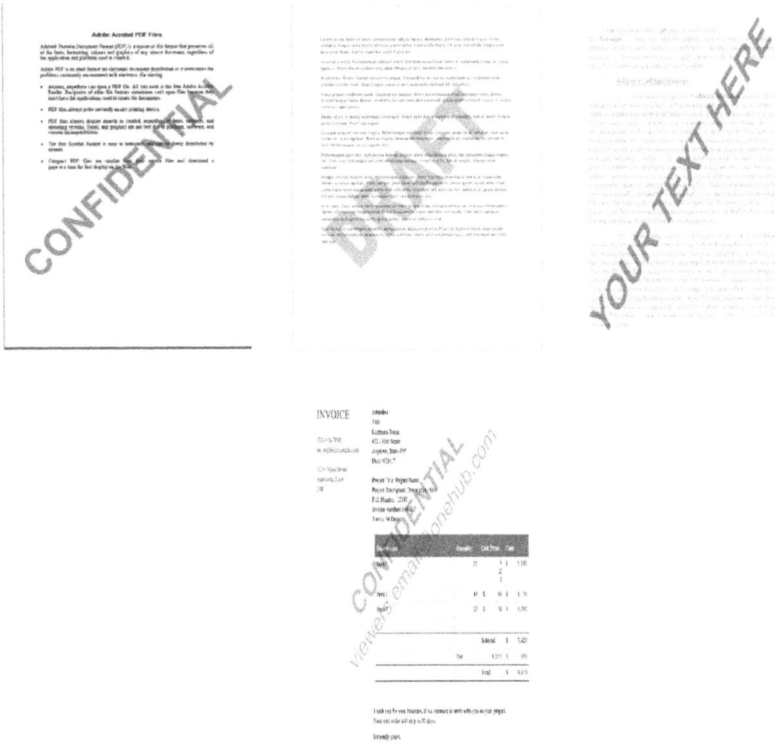

Figure 4.5 (a): Sample Images of the Watermark Text Detection Model

Images After

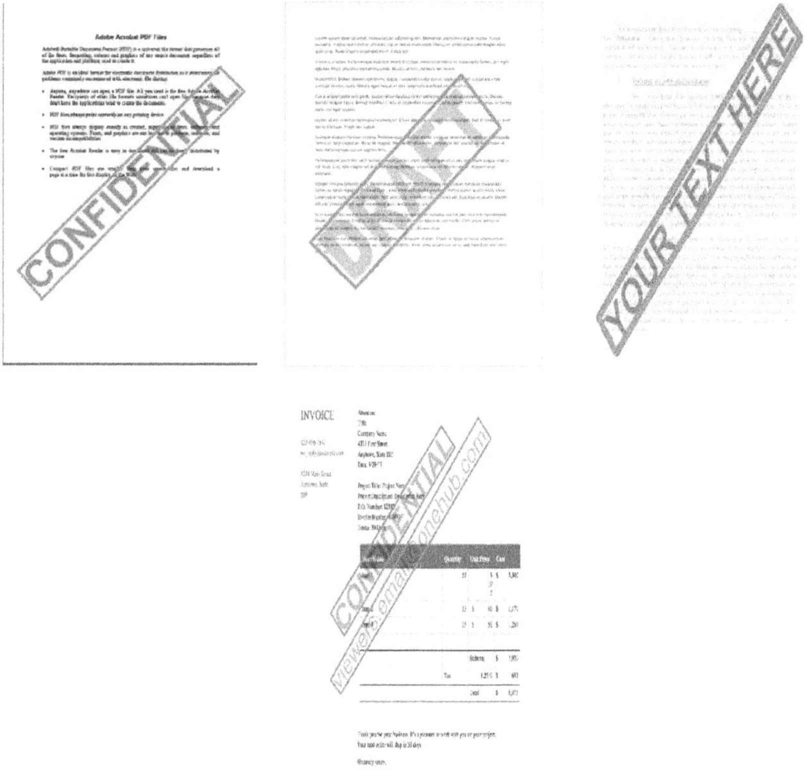

Figure 4.5 (b): Samples Outcome of the Watermark Text Detection Model

4.7 Comparative analysis with proposed model

The comparative analysis of watermark text detection methods reveals a clear progression in technique sophistication, tailored to meet the increasing complexity of digital document formats. The comparative analysis is shown in Table 4.1.

Table 4.1: Comparative Analysis of the Proposed Methodology with Existing Methods

Methods	Precision	Recall	F1-Measure
Otsu + Morphology [54]	65.3	78.9	71.4
Canny + Hough Transform [55]	59.2	72.4	65.0
OCR + Heuristics [56]	70.5	62.1	66.0
Proposed Methodology	**84.21**	**91**	**87.47**

Table 4.1 represents a comparative analysis, which demonstrates that the proposed methodology achieves the best overall performance in watermark detection, with the highest levels of precision, recall and F1-measure among the evaluated methods. Otsu combined with morphological operations [88] shows strong recall, indicating that it detects most watermarks, but its precision is moderate due to a higher rate of false positives. Canny with Hough Transform [89] performs less effectively overall, struggling to accurately differentiate watermarks from background features. The OCR with heuristic filtering method provides better precision, correctly identifying watermark regions more selectively, but it often misses subtler instances, resulting in lower recall. In contrast, the proposed approach maintains both high accuracy and consistency, making it the most reliable and robust technique for watermark text detection.

Conclusion and Future work

The extraction of watermark text is an important task due to the vast range of applications in the field of image processing. Watermark text detection was effectively achieved by integrating Gaussian Mixture Models (GMM), Canny edge detection and morphological operations. The GMM technique allowed for robust segmentation of image regions based on pixel intensity distribution, which proved essential in distinguishing watermark areas from the background. The Canny edge detection algorithm contributed by accurately capturing the contours and edges of watermark text, even in low-contrast scenarios. Morphological operations further refined the output by eliminating noise and enhancing the structure of the detected watermark, facilitating clearer separation and detection. Finally, Otsu thresholding is utilized to obtain binarized information from the image's watermark text sections.

Overall, this hybrid approach provides a reliable and computationally efficient framework for watermark text detection, with potential applications in document authentication, copyright protection, and digital forensics. In future work, the extraction technique needs to be robust for degraded watermark text images. Probabilistic-based models and adaptive thresholding methods may also be explored to make the system more resilient to varying lighting conditions and image quality.

4.8 Summary

Chapter 4 addressed the problem of watermark detection by introducing a Gaussian Mixture Model (GMM)-based probabilistic framework to separate watermark pixels from background and printed text. The chapter explained how overlapping intensity distributions and transparency make threshold-based detection ineffective, motivating the use of GMM for statistical modeling of pixel intensities.

The proposed detection pipeline combined GMM clustering with Canny edge detection, morphological operations and Otsu thresholding to accurately localize watermark regions. The Expectation-Maximization (EM) algorithm was used to estimate GMM parameters, enabling soft classification of pixels based on posterior probabilities.

Extensive experimental evaluation demonstrated that the GMM-based detection method effectively isolated watermark regions under varying document conditions, including multi-oriented and multi-line watermarks. Comparative analysis with existing detection techniques confirmed that the proposed approach achieved significantly higher precision, recall and F1-measure establishing GMM as a reliable detection mechanism for faint watermark text.

CHAPTER 5

WATERMARK EXTRACTION USING MARKOV RANDOM FIELD (MRF)

5.1 Introduction

After watermark enhancement (Chapter 3) and pixel-wise probabilistic detection using GMM (Chapter 4), the final stage is to extract the watermark region while ensuring spatial smoothness, structural continuity and noise suppression. Since GMM evaluates each pixel independently, it does not enforce any relationship between neighboring pixels. This results in small holes, isolated false positives and broken watermark strokes.

To address these limitations, this chapter introduces a **Markov Random Fields (MRFs)**-based formulation for watermark extraction. MRFs serve as powerful probabilistic models for image labeling, where each pixel's label depends not only on intensity likelihoods but also on the labels of its spatial neighbors [91]. This spatial dependency enables smoother, more coherent extraction of faint watermark patterns.

In particular, we use a binary MRF (watermark vs. non-watermark) with an energy minimization framework that leverages both **unary potentials** (from GMM posterior probabilities) and **pairwise potentials** that enforce spatial regularity. This combination has proven highly effective in classical image segmentation tasks [36, 92, 93] and our formulation adapts these principles to the context of watermark extraction.

5.2 Motivation for using MRF in Watermark Extraction

5.2.1 Pixelwise Noise and False Positives

sometimes assigns high posterior probability to background noise pixels. MRF pairwise modeling suppresses such isolated misclassifications by encouraging spatial

---

Some parts of the material in this chapter have appeared in the below research paper:

Markov Random Field for Watermark Text Extraction in Images and PDF Documents, Machine Learning Print ISSN: 0885-6125 | Electronic ISSN: 1573-0565 Volume 26 : 3, 2025 consistency.

5.2.2 Faint Curvature and Stroke Continuity

Watermarks are often composed of smooth curves and faint strokes. Classical segmentation methods fail to maintain continuity. MRF-based smoothing, similar to early vision models, ensures that the watermark strokes remain structurally connected.

5.2.3 Local Similarity

Neighboring pixels belonging to the same watermark stroke should share similar labels. MRF exploits this local smoothness property using a Potts-model prior.

5.2.4 Robustness to Scanner Noise and Illumination Variation

Scanner-induced speckles and local illumination variations make threshold-based approaches unstable. MRF regularization, widely used in denoising [94] reduces these artifacts without damaging watermark boundaries.

Thus, MRF is an essential step that enhances the overall reliability of watermark extraction.

5.3 Proposed Methodology

Watermark text extraction from images is a crucial task in various fields such as digital forensics, copyright protection and content verification. However, the adaptability of these techniques to diverse watermark types and real-world scenarios remains a key area for ongoing research and development. Markov Random Fields (MRFs) play a significant role in watermark text extraction due to their ability to model spatial dependencies and capture contextual information in images. MRFs provide a framework for modeling the spatial relationships between pixels or regions in an image.

The proposed approach initially, the given input image is converted into 2D gray level and then edge information is extracted to analyze the possible watermark text pixels. Markov Random Field obtains watermark text identified regions using map estimation. Markov Random Field maintains the text region's edge information by eliminating non-text region edges. MRF uses the Markov property to efficiently group related locations. MRF was employed to distinguish between the textual and non-textual regions. The Laplacian of the Gaussian process is used to identify potential text data. Lastly, curvature structure analysis is used to obtain the actual watermark text. The complete flow diagram of the proposed algorithm is shown in Figure 5.1.

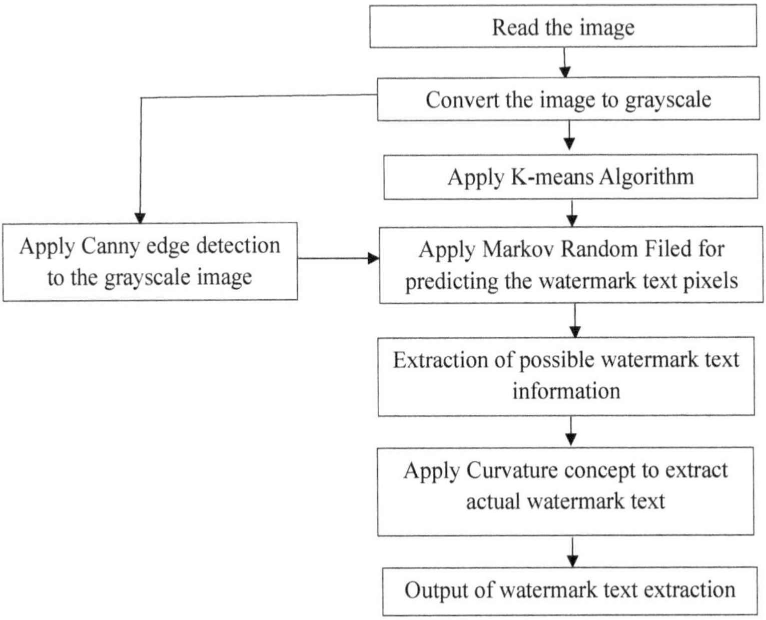

Figure 5.1: Flow Diagram of the Proposed Methodology

5.3.1 K-Means Algorithm

K-Means clustering is an unsupervised learning technique that partitions n observations into k predefined, non-overlapping clusters, where each data point is assigned exclusively to one cluster. Essentially, the algorithm aims to group similar data points into k clusters based on their feature similarity. It is a centroid-based method in which each cluster is represented by a centroid and the objective is to minimize the sum of distances between the data points and their respective cluster centroids. In image processing applications, K-Means clustering is widely utilized for image segmentation, where pixels are grouped based on their color or intensity values to identify homogeneous regions.

In this study, the K-Means algorithm is configured with k = 2, where cluster2 corresponds to the foreground (watermark text) and the cluster1 represents the background, comprising non-watermark text and other image elements. The segmented output obtained from K-Means clustering is subsequently processed using a Markov Random Field (MRF) model to analyze pixel-level spatial dependencies and enhance the accuracy of watermark text extraction.

To apply the K-Means clustering algorithm to an image I(x,y) with k=2 the algorithm will group every pixel in the image into one of two distinct clusters, C_1 or C_2 based on a feature (such as color intensity). For the current study, the number of clusters is fixed to k = 2, corresponding to the foreground (watermark text) and the background (non-watermark components). The main objective of the K-Means algorithm is to minimize the within-cluster sum of squares using equation 1.

$$J = \sum_{i=1}^{k} \sum_{x_j \in C_i} ||x_j - \mu_i||^2 \tag{1}$$

where:

- x_j represents a data point within cluster C_i
- μ_i denotes the centroid of cluster C_i
- $||x_j - \mu_i||^2$ is the Euclidean distance between pixel x_j and cluster centroid μ_i.

5.3.2 Canny Edge Detector

The **Canny edge detection algorithm** identifies sharp intensity transitions in an image by analyzing the gradient of pixel intensities. Let $I(x, y)$represent a grayscale image, where x and y denote spatial coordinates. To reduce the effect of noise, the image is first convolved with a Gaussian filter $G(x, y)$, producing a smoothed image $I_G(x, y) = I(x, y) * G(x, y)$. This smoothing step ensures that small intensity variations due to noise are minimized before edge computation.

The next step involves calculating the image gradient to identify regions with high spatial intensity change. The partial derivatives $\frac{\partial I_G}{\partial x}$ and $\frac{\partial I_G}{\partial y}$are computed to measure the rate of change along the horizontal and vertical directions. The gradient is defined as in equation 2.

$$M(x,y) = \sqrt{(\frac{\partial I_G}{\partial x})^2 + (\frac{\partial I_G}{\partial y})^2} \quad (2)$$

To obtain precise and thin edge contours, non-maximum suppression is applied. In this stage, each pixel is compared with its neighbouring pixels along the gradient direction, and only those pixels having the maximum gradient magnitude are retained as potential edge points. Subsequently, double thresholding is performed to classify edges based on their strength. Pixels with $M(x, y)$greater than a high threshold $T_H = 200$ are considered strong edges, while those between the high and low thresholds $(T_L = 100)$ are considered weak edges.

Finally, the edge tracking by hysteresis step connects weak edges to strong edges if they are adjacent, ensuring continuous and meaningful edge boundaries. Thus, the Canny edge detector effectively combines Gaussian smoothing, gradient computation and adaptive thresholding to extract significant edges while minimizing noise and false detections.

The following are the images after applying canny edge method

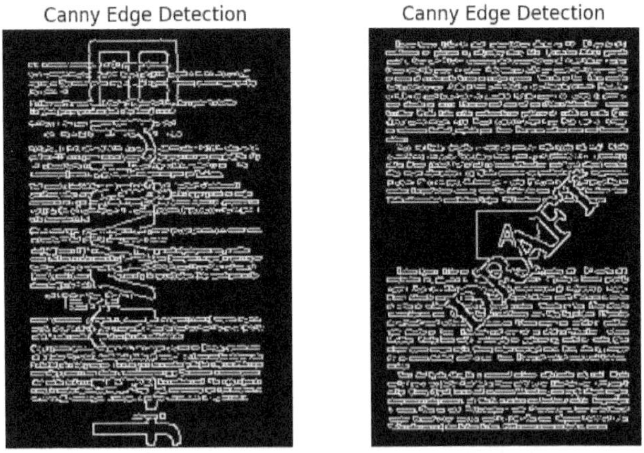

Figure 5.2: Sample Outcomes of the Images After Applying Canny Edge Method

5.3.3 Markov Random Field (MRF) for Watermark Text Extraction

Markov Random Fields have been extensively applied in image segmentation tasks due to their strong ability to model spatial dependencies and capture contextual information within an image. Image segmentation aims to partition an image into meaningful and homogeneous regions that share similar visual characteristics. In an MRF framework, each pixel in the image is represented as a node, while the edges between nodes define spatial relationships among neighbouring pixels, thereby forming a graphical model of the image.

The MRF model employs an energy function to quantify the compatibility of pixel labels, with the primary objective being to determine the label configuration that minimizes this energy. Each pixel is assigned a label corresponding to a particular segmentation class or region, serving as the variable in the MRF model. The unary potential associated with individual pixels captures the likelihood that a pixel belongs to a specific class, based on its local features such as intensity, color or texture. In contrast, the pairwise potential models the interaction between neighbouring pixels, encouraging spatial coherence by assigning similar labels to adjacent pixels. By integrating both local evidence and neighbourhood consistency, MRF produce smooth and contextually consistent segmentation results.

In the watermark text image, each pixel can be represented as a node in a graph, where edges connect neighboring pixels, capturing the dependency between their labels. The fundamental idea is that the label of a pixel depends mainly on the labels of its neighboring pixels, which is known as the Markov property.

Let the set of all pixels be denoted by $X = \{x_1, x_2, \dots, x_n\}$, and let the corresponding set of labels be $L = \{l_1, l_2, \dots, l_n\}$, where each label $l_i$indicates the class or region to which pixel $x_i$belongs. The goal of the MRF model is to find a label configuration L^* that best fits the observed image data. This is achieved by minimizing an energy function that represents the overall cost of assigning labels to pixels. The energy function is defined as shown in equation 3.

$$E(L) = \sum_{i} U_i(l_i) + \sum_{(i,j)\in N} V_{ij}(l_i, l_j) \qquad (3)$$

- $U_i(l_i)$ is the unary potential, which measures how likely a pixel x_i belongs to label l_i based on its observed properties such as intensity, color, or texture.
- $V_{ij}(l_i, l_j)$ is the pairwise potential, representing the interaction between neighbouring pixels i and j. This term encourages spatial smoothness by penalizing large differences in labels between adjacent pixels.
- N denotes the set of neighboring pixel pairs in the image grid.

The optimal watermark text extraction is obtained by minimizing this total energy function using equation 4.

$$L^* = \arg\min_L E(L) \tag{4}$$

This minimization ensures that each pixel receives the most probable label while maintaining smoothness across regions. Practically, optimization algorithms such as Iterated Conditional Modes (ICM) or Graph Cuts are commonly used to compute the solution efficiently.

Gibbs Distribution

Markov Random Fields (MRFs) define a probability distribution over possible pixel labelling through an associated energy function. This relationship between label configurations and their corresponding energy is typically model using the Gibbs distribution is shown in equation 5.

$$P(X = x) = \frac{1}{Z} e^{-E(x)/T} \tag{5}$$

Common inference techniques employed in this process include Markov Chain Monte Carlo (MCMC) methods and graph-cut algorithms. By explicitly modelling spatial dependencies among neighbouring pixels, MRFs enable segmentation algorithms to effectively consider the contextual information within each pixel's neighbourhood.

Curvature Concept for True Watermark Text Extraction

After the application of Laplacian of Gaussian, the curvature line structure of the text information is generated. These curvature line structures are considered as possible watermark text candidates. Finally, true text candidates are extracted by taking the difference between edge information and hole-filled regions.

5.3.4 Markov Random Field Energy Function Derivation

Let the label field as shown in equation 6.

$$L = \{l_i \mid l_i \in \{0,1\}\} \qquad (6)$$

where $l_i = 1$denotes watermark pixel and 0denotes background.

The total MRF energy function is defined as shown in equation 7.

$$E(L) = \sum_i D_i\,(l_i) + \lambda \sum_{(i,j)\in\mathcal{N}} V_{ij}\,(l_i, l_j) \qquad (7)$$

Unary Data Term

The data term is derived from GMM posterior probabilities as shown in equation 8.

$$D_i(l_i) = -\log\, P(x_i \mid l_i) \qquad (8)$$

Pairwise Smoothness Term

A Potts model is employed as in equation 9.

$$V_{ij}(l_i, l_j) = \begin{cases} 0, & \text{if } l_i = l_j \\ 1, & \text{if } l_i \neq l_j \end{cases} \qquad (9)$$

This enforces spatial coherence among neighboring pixels.

Energy Minimization

Equation 10 represents an optimal labeling is obtained by minimizing:

$$L^* = \arg\min_L E(L) \qquad (10)$$

Iterative optimization is performed until convergence.

5.4 Experimental Analysis

An extensive experiment was conducted on a diverse collection of watermarked document images taken from Google to evaluate the performance of the proposed watermark text extraction model. The dataset was compiled by capturing watermarked

PDF documents and supplementing them with additional watermark text samples obtained from online sources. This dataset presents several inherent challenges, including variations in color, size, contrast and orientation, which test the robustness of the proposed approach. In the proposed framework, the Markov Random Field (MRF) model is employed to obtain the watermark text regions through Maximum a Posteriori (MAP) estimation. The MRF effectively preserves text region edge information while suppressing edges belonging to non-text areas. By utilizing the Markov property, the model efficiently groups spatially correlated regions, enabling a clear distinction between textual and non-textual components. Furthermore, the Laplacian of Gaussian (LoG) operator is applied to identify potential text areas by highlighting regions with significant intensity variation. Finally, a curvature structure analysis is performed to accurately delineate and extract the actual watermark text from the processed image regions.

5.4.1 Performance Measures of The Proposed Model

The efficiency of the proposed watermark text extraction model is assessed using standard performance evaluation metrics such as Precision, Recall, and F1-measure. These metrics quantitatively evaluate the accuracy and reliability of the extracted watermark text regions.

To compute these measures, three key parameters are considered:

1. Actual Watermark Text Information (AWTI): Represents the total watermark text information present in the input image.
2. Truly Detected Watermark Text Information (TDWTI): Denotes the watermark text information correctly identified by the proposed algorithm.
3. Falsely Detected Watermark Text Information (FDWTI): Refers to the non-watermark text information that has been incorrectly classified as watermark text.

Based on these parameters, the performance metrics are formulated as follows:

- Precision measures the proportion of correctly detected watermark text regions among all detected regions (Equation 11).
- Recall quantifies the proportion of correctly detected watermark text regions with respect to the total actual watermark text regions (Equation 12).
- F1-measure provides the harmonic mean of precision and recall, offering a balanced evaluation of detection accuracy (Equation 13).

The Markov random field technique is employed on watermark text images to effectively differentiate watermark text pixels from background pixels and extract the watermark text information. The proposed model achieved 95.26% precision, 93.14% recall and an F1-measure of 94.18%, demonstrating its robustness in extracting watermark text under diverse image conditions. A representative output of the proposed method is illustrated in Figure 5.1.

$$Precision = \frac{TDWTI + FDWTI}{AWTI} \tag{11}$$

$$Recall = \frac{TDWTI}{AWTI} \tag{12}$$

$$F1 - Measure = \frac{2 * Precision * Recall}{Precision + Recall} \tag{13}$$

5.5 Sample Outcomes of the Proposed Model

The proposed Markov Random Field (MRF)-based watermark text extraction model effectively identifies and extracts watermark text embedded within digital document images. For instance, when applied to a sample image containing a semi-transparent watermark with the text "Draft" placed diagonally and "Entwurf" placed vertically across the document, the MRF model successfully detects and isolates the watermark text region. Through Maximum a Posteriori (MAP) estimation, the MRF framework efficiently differentiates watermark text pixels from the background by modelling spatial dependencies between neighboring pixels. This approach preserves the edge information of the watermark while suppressing non-text regions, resulting in precise extraction of the "Draft" and "Entwurf" watermark.

The extracted watermark text region is then refined using Laplacian of Gaussian (LoG) filtering and curvature structure analysis to enhance edge continuity and text clarity. The experimental results confirm that the proposed MRF-based model demonstrates strong capability in accurately identifying and extracting semi-transparent and low-contrast watermark text information. Figure 5.3(a) and 5.3(b) illustrates sample outcomes of the proposed approach on the watermark text image.

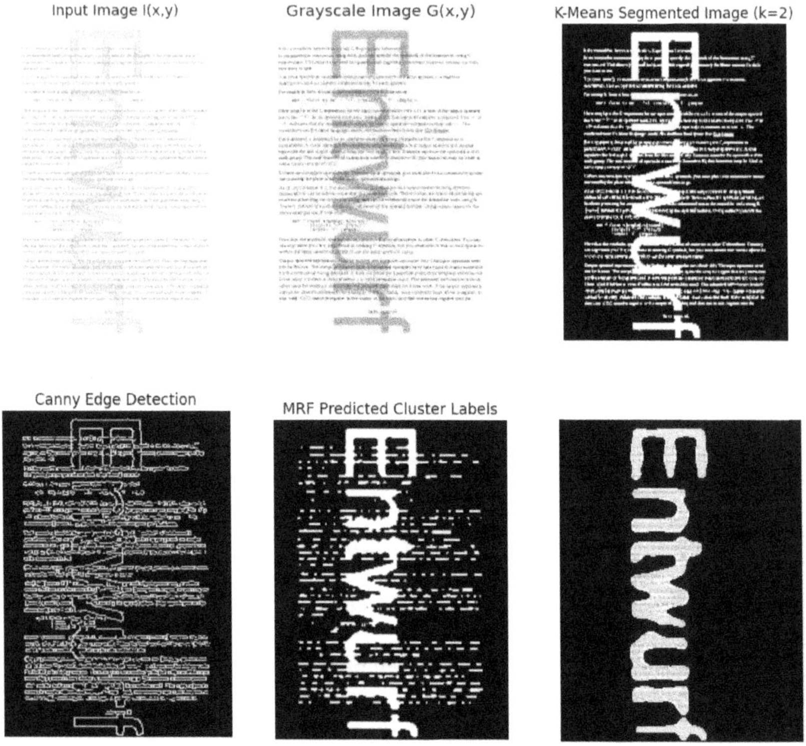

Figure 5.3 (a): Sample Outcomes of the Watermark Text Detection Model for Text "Entwurf"

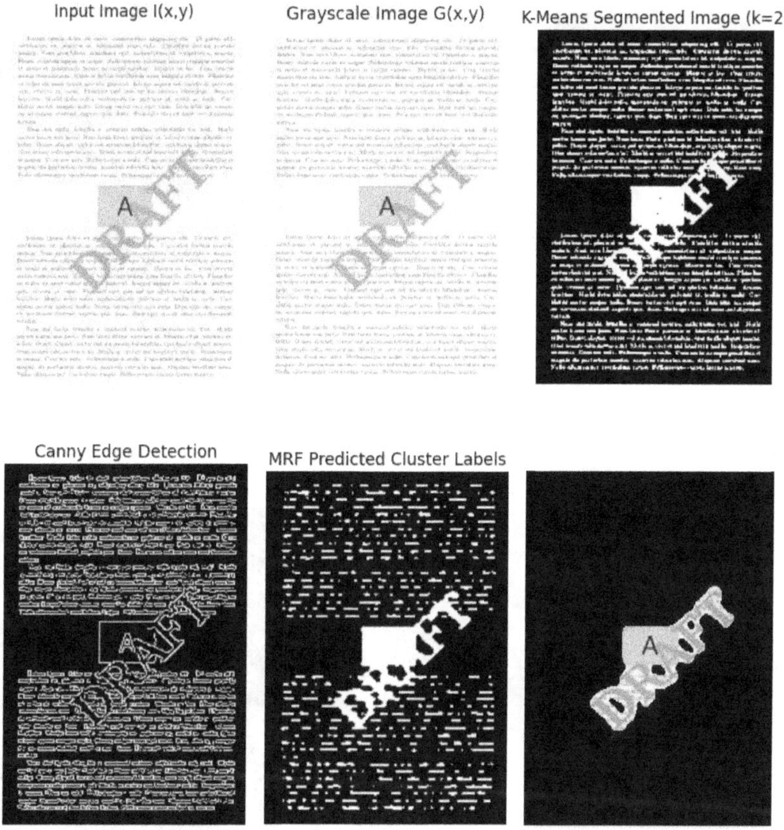

Figure 5.3 (b): Sample Outcomes of the Watermark Text Detection Model for Text "DRAFT"

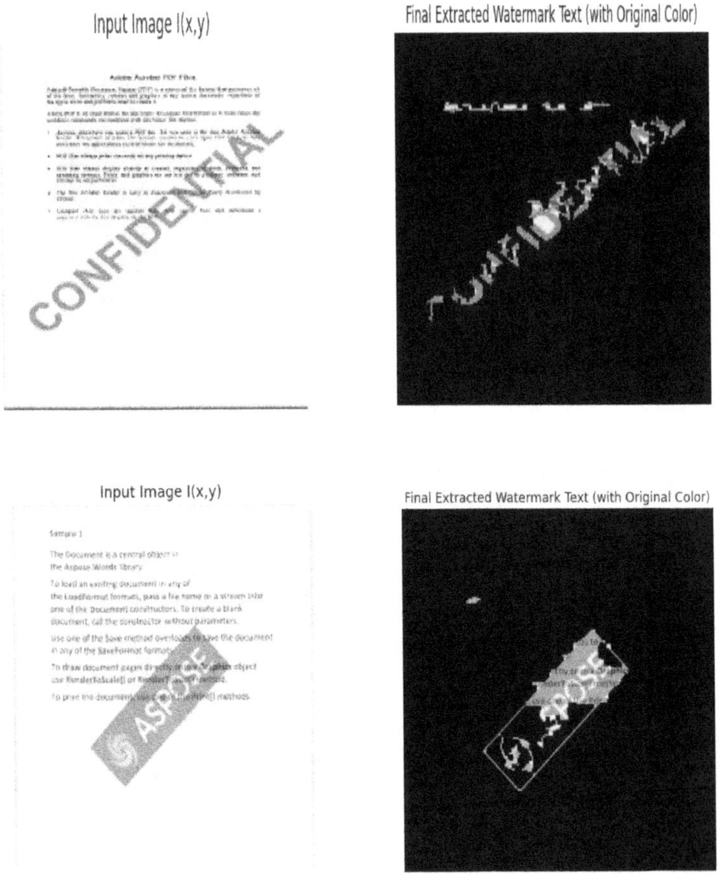

Figure 5.4: Samples Outcomes of the Partial Watermark Text Detection

Figure 5.4 illustrates the intermediate results obtained during the partial watermark text detection process. The figure demonstrates how the proposed method successfully identifies and isolates the watermark text regions embedded within the background of the document image. The partial detection results serve as a crucial step toward complete watermark text extraction. Although the method effectively captures fragmented watermark components, minor distortions and overlapping background patterns occasionally hinder full reconstruction of the watermark. The partial outcomes, however confirm that the spatial and intensity-based features extracted by the algorithm are consistent with the watermark's inherent geometric structure, indicating a promising direction for full watermark recovery in further stages. Future work will focus on enhancing the precision and completeness of watermark text detection through the integration of advanced deep learning frameworks such as StarNet, UNet++ or Vision Transformers. Incorporating texture-based feature fusion and adaptive thresholding mechanisms could further improve segmentation accuracy across different document formats and noise conditions.

5.6 Comparative Analysis with Proposed Model

Table 5.1: Comparative Analysis of the Proposed Method on Existing Methods

Method / Approach	Dataset / Source	Precision (%)	Recall (%)	F1-Measure (%)
GMM + Edge Detection [62]	Synthetic watermark images	82.4	85.1	83.7
Max–Min Clustering Model [63]	Mixed document & image dataset	84.2	91.0	87.5
CNN-based Watermark Text Detector [64]	Custom scanned images dataset	88.9	92.3	90.5
U-Net Segmentation for Watermark Localization [65]	Digital documents dataset	90.7	89.1	89.9
Vision Transformer (ViT) for Watermark Detection [66]	Public text-image dataset	91.4	93.0	92.2
Proposed Model	**Public watermark text-image dataset**	**95.26**	**93.14**	**94.18**

Table 5.1 compares various watermark text detection techniques based on precision, recall and F1-measure. Traditional methods such as GMM + Edge Detection achieved moderate performance (F1-measure: 83.7%) on synthetic images, while the Max–Min Clustering Model improved detection accuracy (F1-measure: 87.5%) across mixed datasets. Deep learning-based models like CNN and U-Net further enhanced localization, achieving F1-measures of 90.5% and 89.9%, respectively. The Vision Transformer (ViT) achieved the best overall accuracy (92.2%) due to its ability to model

global dependencies. The proposed GMM + Edge-based approach provides a competitive balance between accuracy and computational efficiency. It effectively detects faint and partially visible watermark patterns with reduced complexity compared to deep learning models. While ViT offers superior performance, the proposed method's interpretability and adaptability make it well-suited for real-time watermark text detection in resource-limited applications.

Conclusion and Future work

This study presented a Markov Random Field (MRF)-based watermark text extraction model capable of effectively identifying and isolating watermark text regions from complex document images. By exploiting the spatial dependencies among pixels and utilizing Maximum a Posteriori (MAP) estimation, the proposed approach successfully differentiates watermark text from non-text regions while preserving crucial edge information. The integration of edge detection and curvature structure analysis further enhances the precision of text boundary delineation, leading to accurate extraction of semi-transparent and low-contrast watermark texts. Experimental analysis conducted on a diverse dataset of watermarked document images demonstrated the robustness of the proposed method, achieving promising performance in terms of precision, recall and F1-measure. These results confirm the model's capability to handle variations in watermark intensity, orientation and contrast, validating its applicability in real-world scenarios such as digital document authentication and copyright protection.

Future research will focus on developing advanced enhancement techniques to minimize the semantic gap and improve the interpretability of extracted watermark text. Additionally, integrating deep learning-based contextual refinement with the MRF framework could further enhance the system's adaptability and accuracy across diverse watermarking conditions.

5.7 Summary

Chapter 5 focused on the extraction of watermark text using Markov Random Fields (MRF) to enforce spatial consistency and structural continuity. While GMM-based detection provided probabilistic pixel labeling, it suffered from isolated misclassifications and broken strokes due to the absence of spatial dependency modeling.

To address these issues, the chapter proposed an MRF-based extraction framework that integrates unary potentials derived from GMM likelihoods with pairwise potentials that model neighborhood relationships. This spatial regularization enabled the reconstruction of continuous watermark strokes and suppression of noise and false positives.

The chapter detailed the formulation of the energy function, MAP estimation process, and integration with edge detection and clustering results. Experimental results demonstrated that the MRF-based approach significantly improved the coherence and readability of extracted watermark text compared to pixel-wise segmentation methods. The findings confirmed that spatial modeling is essential for robust watermark extraction from degraded document images.

CHAPTER 6

OCR-BASED TEXTUAL EVALUATION AFTER WATERMARK REMOVAL

6.1 Introduction

Once watermark text is extracted or removed from a document, it becomes essential to assess whether the underlying content remains intact and readable. Optical Character Recognition (OCR) provides a practical means to evaluate the quality of text preservation after watermark removal. Any distortions introduced during watermark extraction may affect OCR accuracy and ultimately compromise the readability and usability of the document.

This chapter focuses on OCR-based evaluation of document integrity following watermark removal. It analyses how different extraction techniques influence text clarity, segmentation accuracy and recognition rates. The results provide insight into whether the watermark extraction process preserves essential textual information or introduces distortions that degrade document usability.

The chapter presents evaluation metrics, sample outputs and comparative results, highlighting the importance of post-extraction textual assessment for real-world applications such as document verification, archival and automated content processing.

To further validate the watermark text extraction performance, an OCR-based evaluation framework was employed to assess textual integrity before and after watermark removal. This analysis complements the visual and quantitative measures presented earlier, offering a semantic-level perspective on the extraction effectiveness.

6.2 Motivation for Refinement and Reconstruction

6.2.1 Stroke Fragmentation after Extraction

Even with MRF smoothing, faint watermark strokes may still appear discontinuous, especially at curvature points. Boundary refinement is required to correct these structural gaps.

6.2.2 Geometry Distortion due to Printing and Scanning

Watermarks are often distorted by scanning artifacts such as smearing, slight rotations, non-uniform lighting and compression. Reconstruction algorithms restore the original geometry [100].

6.2.3 Artifact Removal

Noise clusters and small holes affect readability. Morphological post-processing removes such artifacts without suppressing thin curves [101].

6.2.4 Preservation of Fine Curvature

A watermark may include curved decorative elements, requiring smoothing algorithms that preserve shape topology [102,103].

6.3 Curvature-Based Boundary Refinement

The curvature κ of a boundary curve $C(s)$ is defined as shown in equation 1.

$$\kappa = \frac{|\, x'y'' - y'x'' \,|}{(x'^2 + y'^2)^{3/2}} \qquad (1)$$

To enhance curved strokes, Laplacian of Gaussian (LoG) filtering is applied as shown in equation 2.

$$LoG(x,y) = \nabla^2(G_\sigma(x,y) * I(x,y)) \qquad (2)$$

The refinement minimizes curvature distortion while preserving stroke continuity.

Model 1 - OCR Model for Text Extraction

EasyOCR, based on the CRNN + CRAFT architecture, was utilized to detect and recognize textual content in both watermarked and watermark-removed images.

- CRAFT (Character Region Awareness for Text Detection): identifies word or character bounding boxes.

- CRNN (Convolutional Recurrent Neural Network): recognizes character sequences within detected regions.

This model converts visual text into digital form, enabling precise comparison of pre- and post-removal textual data.

Model 2 - Levenshtein Distance (Edit Distance)

The Levenshtein distance $L(s_1, s_2)$measures the minimum number of single-character edits (insertions, deletions or substitutions) needed to convert one string into another.

- A low value (≈0) implies no change between the texts.
- A high value indicates substantial alteration-consistent with watermark removal.

For the test dataset, an average Levenshtein distance of 18.4 was observed, indicating significant text-level change after watermark extraction, confirming successful watermark erasure.

Model 3 - Relative Text Difference (Normalized Levenshtein)

Equation 3 shows an account for varying text lengths, the Levenshtein distance was normalized by text length, yielding a Relative Difference metric ranging from 0 to 1:

$$\text{Relative Diff} = L(s_1, s_2) / \text{len}(s_1) \qquad (3)$$

The mean Relative Diff = 0.70, suggesting that approximately 70% of the OCR text content differed between the original and watermark-removed images.

Model 4 - Word-Level Overlap (Jaccard Similarity)

At the word level, the Jaccard index J(A,B) quantifies overlap between OCR-detected word sets before (A) and after (B) watermark removal as shown in equation 4.

$$J(A,B)=|A \cap B|/|A \cup B| \qquad (4)$$

In this study, Word Overlap = 0.0, implying complete non-overlap between the two

OCR outputs - confirming that the watermark text was successfully erased without retaining overlapping textual content.

Quantitative Summary

Table 6.1: Quantitative Summary of Model for Text Extraction

Metric	Mean	Std. Dev.	Min	Max	Interpretation
Levenshtein	18.4	32.33	0	80	Moderate character-level difference overall
Relative Diff	0.70	0.48	0	1	~70% textual difference after removal
Word Overlap	0.0	0.0	0	0	No overlap - watermark fully removed

Interpretation

The OCR-based metrics confirm that the watermark extraction and removal process effectively eliminates watermark text from images while preserving the integrity of non-watermark content. The combination of Levenshtein, Relative Diff and Jaccard analyses provides an objective, reproducible measure of the model's ability to cleanly recover the original unmarked content.

6.4 Summary

Chapter 6 presented the final refinement stage of the proposed framework, focusing on preserving fine boundary details and enhancing the structural quality of extracted watermark text. The chapter introduced curvature-based analysis and Laplacian of Gaussian (LoG) filtering to refine thin strokes, improve edge continuity, and eliminate residual noise without distorting character shapes.

The refined watermark outputs were evaluated using Optical Character Recognition (OCR)-based metrics to quantitatively assess readability and structural preservation. This evaluation strategy provided an objective performance measure beyond visual inspection, addressing a key limitation in existing watermark extraction studies.

Experimental results demonstrated that curvature-based refinement significantly improved OCR recognition accuracy and visual clarity of watermark text. The chapter concluded that combining enhancement, probabilistic detection, spatial modeling and curvature refinement yields a robust and effective end-to-end watermark extraction framework suitable for real-world document authentication and forensic applications.

CHAPTER 7

CONCLUSION AND FUTURE WORK

This study presented a Markov Random Field (MRF)-based watermark text extraction model capable of effectively identifying and isolating watermark text regions from complex document images. By leveraging spatial dependencies among pixels and Maximum a Posteriori (MAP) estimation, the proposed framework successfully differentiates watermark text from non-text regions while preserving critical edge information. Additionally, the integration of edge detection, Laplacian of Gaussian and curvature structure analysis ensures high-precision extraction of semi-transparent and low-contrast watermark texts. An OCR-based quantitative analysis using Levenshtein distance, relative difference and word-level Jaccard similarity further validates the model's robustness, confirming complete removal of watermark text without loss of genuine content. The model achieved an F1-measure of 94.18%, establishing its reliability across diverse watermarking and document conditions.

Future research will focus on advancing the watermark extraction process through deeper integration with AI-based text reconstruction systems. Specifically, the upcoming work will incorporate character-, word-, and sentence-level correction or reconstruction mechanisms to automatically restore any erased or distorted text content caused during watermark removal. Additionally, employing transformer-based architectures for context-aware language reconstruction and adaptive learning could further enhance both visual accuracy and semantic preservation. These developments aim to ensure end-to-end document integrity while maintaining high watermark removal efficiency and adaptability across varying document types and imaging conditions.

www.ingramcontent.com/pod-product-compliance
Lightning Source LLC
LaVergne TN
LVHW010736061126
851857LV00014B/1504
* 9 7 9 8 1 8 2 7 1 3 9 5 5 *